THE ATOM

how it really works

By

Keith Dixon-Roche

(based upon work by Isaac Newton

and Charles-Augustin de Coulomb)

Keith Dixon-Roche © 2017 to 2019

THE ATOM

how it really works

Published by CalQlata
info@CalQlata.com

First published August 2019
Copyright © Keith Dixon-Roche 2019

Contents

Preface

Following the completion of my work on Newton's laws of orbital motion, which surprisingly (for me) encompasses the atom, I decided that it would be appropriate to publish a book that concentrates on the atom "according to Newton"; my apologies, "and Coulomb".

The atomic model presented here is based upon the work carried out by scientists well before the start of the twentieth century (refer to Appendix A6). I.e. this model of the atom could – and should - have been established long before we were all perplexed with quantum theory.

The model described here not only works in theory, it can also be seen to work in practice. It realistically describes the universe and all the matter within, and not once has it been necessary to claim that "the rules of physics do not apply".

In fact, using this model, it has been possible to explain all the properties of every Quanta in every element at any temperature, something that quantum theory only ever claimed for *the velocity of a hydrogen atom* at what was proposed as rest-mass; a condition that has yet to be explained.

Having discovered that just as Einstein never derived the formula $E=mc^2$ (it was Henri Poincaré), Bohr never derived the 'Bohr' radius (a_o); it was originally derived by Johannes Rydberg. <u>In my opinion</u>, this is little better than plagiarism, and should be corrected, even if it is a little late.

To the best of my knowledge, in this publication, I have tried to attribute all terms and discoveries to their original proposers. If I have misrepresented anybody, I apologise in advance. Please let me know and it will be corrected in a subsequent edition.

The theories presented here will ultimately form the basis for the mathematical theory of "*The Molecule*" – a future publication – which will form the basis for a molecular calculator; the only way forward for perfect chemistry (chemicals, materials & medicines) that can be designed in minutes from a computer terminal (no need for laboratories).

I consider this publication to be the precursor to what will be the most important discovery for life on this planet; *the molecular calculator*.

Keith Dixon-Roche 2019

The following is a statement by me (Keith Dixon-Roche) with regard to the authorship of this work:

I do not claim to have discovered the real atom all by myself; it was possible only because of the work carried out previously by
Isaac Newton and Charles-Augustin de Coulomb.

I believe it only right and just that <u>all</u> contributors should be rightfully acknowledged in publication authorship and I regard both of these people as important co-authors and contributors to its discovery.

I need to make it absolutely clear that despite trying to give these contributors the recognition they deserve:

Amazon Books denied them this recognition (in writing), which is the reason they are not included as co-authors on the cover of this publication.

1 Introduction

Newton's laws of orbital motion show us that the atom is not the unpredictable, fuzzy, indeterminate structure described by quantum theory; it is in fact similar to all the other universal orbital systems as described by Isaac Newton. The only differences being that atomic satellites (electrons) orbit in circular paths (because they provide their own kinetic energy) and also obey the force laws defined by Charles-Augustin de Coulomb and William Gilbert.

The atom is a stunningly brilliant but simple structure that has no need for sub-atomic particles, uncertainty, unpredictability or statistical behaviour. In all my work on Newton's orbits, the atom is by far the most amazing; it is the only feature of nature that, perhaps, could entice me to believe that somewhere out there, exists a genuine genius.

The entire universe comprises a huge number (>2.8E+75) of only two particles (protons and electrons). Every proton is exactly the same as every other proton and every electron is exactly the same as every other electron. Protons and electrons exist as proton-electron pairs; a stationary proton orbited by a moving electron. Apart from a few lone protons (H+) and free-flying electrons, there is nothing else out there.

Neutrons are simply proton-electron pairs that have united as a result of the orbiting electron having achieved the *speed of light.* They can only be created, and can only exist, inside an atom. And they decay into their component parts, a proton and an electron as alpha & beta-particles (respectively) when ejected.

Whilst neutrons are not particles in their own right, they are exceptionally important; they are the universe's energy storehouse. They provide all the energy necessary to ignite the next *Big-Bang.*

Refer to my publication "The Life & Times of the Neutron – *universal energy*" to understand how they can also solve all mankind's energy problems at a stroke.

The atom is such a simple system, it is difficult to see how it can create such a diverse universe, but it does and that is its genius. These two particles are responsible for creating everything in the universe from galactic force-centres to stars, planets, the *Big-Bang* and life itself. And they recreate each new universal period with no outside help.

The atom is the universe's micro-orbital system, involving self-propelled satellites (electrons) in circular orbits.

Everything else in the universe operates as a macro-orbital system (elliptical orbits). Refer to my publication "The Universe – *how it really works*".

Unresolved issues are highlighted in the text with the superscript [?] in which '?' will be replaced with a number that can be found in Chapter 8

1.1   What can we do with this knowledge?

If we know how the atom works, we can derive a working model for the molecule.

With a mathematical model for the molecule, we can create a molecular calculator.

With a molecular calculator, we can solve all chemical, including medical and material, problems in minutes from a computer terminal with semi-skilled technicians and no trial and error (laboratory experimentation).

National and international health authorities will no longer need to enforce decades of *real-life* tests to verify a medicinal product. Every chemical will be 100% correct with no unforeseen side-effects.

Pharmaceutical companies and hospitals will no longer be able to define a patient's right to live or die based upon their ability to pay.

1.2   How This Book Is Organised

This book comprises 8 sections, the first four of which provide similar information but in a different form:

2 Narrative

A written description that gives a general overview of the various discoveries made in this book. It is devoid of formulas and mathematical complexity with a view to providing a *'light-read'*!

3 Calculation Procedures

A compilation of the mathematical formulas supporting the narrative, including how to use them. This section has been written to simplify their use.

4 Calculation Results

A collection of [mostly] tabulated calculation results for selected examples using the formulas provided in section 3 (above).

5 Physical Constants

All the physical constants (including electrical properties such as Volts, Amps, Henries, Farads, Ohms, etc.) are provided (to ≤15 decimal places) in terms of the same four basic units; length, time, mass and charge and two ratios: m_e, e, R_n, t_n & ξ_υ, ξ_m

6 Support

The laws of thermodynamics.

7 Things You Can Do!

A list of unresolved issues.

Appendices

References, symbols, glossary, etc. used throughout this book.

2 The Narrative

A written description that gives a general overview of the various discoveries made in this book. It is devoid of formulas and mathematical complexity with a view to providing a *'light-read'*!

The Atom

2.1   Energy

Energy is the only constituent in our universe. It holds the universe together and makes it work; and it comes in various forms:

Electrical (charge & field)

Magnetic (charge & field)

Electro-magnetic (energy transfer)

Potential (static)

Kinetic (dynamic)

What Newton originally referred to as his "conservation of momentum" was the forerunner of what we today call the "conservation of energy", and has since become a fundamental law of natural physics. It has now become the first law of thermodynamics (refer to Chapter 6).

Both particles (proton and electron) possess an electric-magnetic charge, that generates the electro-magnetic and field energies that power the universe.

Electricity and magnetism work contrary to each other:

Magnetic charge is **constant** in both particles and **accrued** between all others (in the universe). It travels **from positive towards negative**.

Electrical energy is constant in the electron and **variable** in the proton, and **shared** between all others (in the universe). It travels **from negative towards positive**.

Electro-magnetic radiation (heat & light) is the means whereby energy is transferred between proton-electron pairs as shown in Figs 1 & 2.

Potential energy, which exists between all particles, is the attraction (negative) or repulsion (positive) of either electrical or magnetic charge.

Kinetic energy, which exists in all moving particles, is always positive and transferred via electrical, magnetic, electro-magnetic or impact [potential] energy.

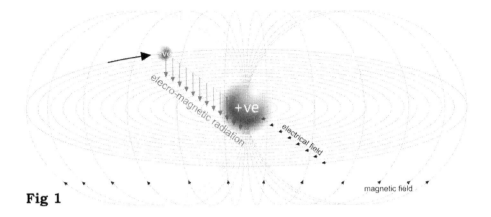

Fig 1

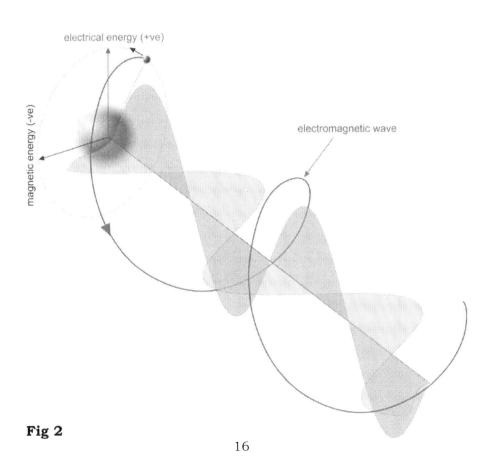

Fig 2

2.1.1 Heat & Light

It is important to remember that all the electro-magnetic energy generated in the universe is just that; electro-magnetic energy (EME). It possesses; no light, temperature or sound – nothing, apart from energy.

If you or I, devoid of electrons – impossible I know, but bear with me – were to sit in the space between the sun and the earth, we would not be able to detect the sun's radiated EME. It would be invisible in every sense to the fictitious you (or me). EME is useless to all forms of life unless it can be detected.

Whilst EME doesn't deteriorate with distance travelled, we don't feel the sun's surface temperature (5778K) here on Earth because the energy *density* (Joules per square metre) radiated at the sun's surface is distributed over a spherical surface area between 45,000 and 48,000 times greater (dependent on the time of year). Therefore, the EME *density* we receive will be correspondingly less.

Life here on Earth, has evolved to detect and use this energy through our complex molecules. The trouble is, such molecules have energy tolerance levels, outside which they would no longer function; i.e. their state-of-matter strength or condition (gas-viscous) could change, or inter-atomic bonding could fail.

For example; if a block of viscous iron received sufficient EME to increase its proton electrical charge energy above that of its atomic magnetic field energy, it would become a gas. And it would cease to be *a block of iron*.

There are tolerance levels regarding acceptable amounts of EME any living organism can receive and remain functional. Therefore, all living organisms have developed senses, that can be used to ensure that these tolerance levels are maintained.

We (humans) have five senses – if you exclude time – smell, touch, taste, sight and hearing; each of which were developed for application and protection.

2.1.1.1 Heat

Temperature is the EME emitted by the proton-electron pairs whose electrons are orbiting in the innermost atomic shell.
Heat is the kinetic energy in an atom's electrons. The heat we *feel* is from the senses we have developed to tell us when this kinetic energy is too high or too low. Electron kinetic energy is generated by the EME it absorbs from its surroundings.

You can't damage a block of iron, so it doesn't need senses. It doesn't matter how many times you change it from gas to viscous and back again it always returns to iron. The higher its temperature the stronger its atoms remain, until the innermost electrons achieve the *speed of light*, when they will become a different element (Z-1 or Z-2).

All the EME in our environment is shared between all of our electrons. The greater the EME *density*, the greater the *heat* we feel. Irrespective of the *temperature* of the atoms that generated the EME, if the energy *density* can be shared throughout all the electrons in our body without exceeding its tolerance levels, we will remain functional.

In other words; it is not the *temperature* of the atoms emitting the EME that defines our body-temperature; it is defined by the quantity of EME absorbed by our body's electrons.

The highest possible temperature that can be generated naturally in our universe is 623316124.717178 K. This is the temperature of a proton-electron pair immediately before it becomes a neutron.

All Heat Is Radiated:

Conduction *is the transfer of EME between the electrons in adjacent atoms in viscous matter.*

Convection *is the movement of gaseous atoms (and molecules) to a position where the repulsive electrical charge energy in their protons can balance in three-dimensions, and where it can transfer this energy to atoms with less. It only occurs in gases that are under the influence of potential energy.*

2.1.1.2 Light

Question: *"If colour is defined by EME, and the surface of the sun is at a temperature of 5778K and looks yellow; how can my towel (at 300K) also look yellow?"*

Not an easy question to answer, but I'll try …

Colour is a range of EME wavelengths (4E-07m > 8E-07m) that we cannot detect until it is absorbed by the electrons in our optical receptors (eyes). **Light** is simply EME intensity, or put simply; the number of electro-magnetic rays per square metre.

We (humans) have developed eyes to detect a particular bandwidth of EME that best suits our purpose. Other lifeforms have developed receptors that best suit their own environment which may be outside the aforementioned *optical* range. Irrespective of a lifeform's preferred optical bandwidth, the purpose of sight is the same; to see what's in its environment and how best to exploit it.

Every proton-electron pair in the universe, *including that block of wood in the garden*, emits EME at a wavelength commensurate with the kinetic energy in its electrons. But you cannot see the EME radiated by that block of wood here on Earth, because it is radiated in the infra-red range. This is the reason infra-red cameras reveal objects in the dark here on Earth. They are actually capturing the electro-magnetic energy given off by the objects themselves, not the electro-magnetic energy radiated by the sun.

Unlike in a prism, the diffraction of light through natural matter is not organised. The image of that block of wood, is the sun's EME reflected and/or refracted by or through its constituent atoms and molecules.

Just as with heat, if the EME received by your eyes is not too intense, i.e. its density remains within your body's tolerance levels, the sun's rays you see will do you no harm.

Whilst our sun's *surface* comprises proton-electron pairs as hydrogen and helium due to its heat, its internals comprise all of nature's elements. It therefore emits a preponderance of electro-magnetic radiation at all the wavelengths from yellow to gamma. All the infra-red, micro-wave and radio-wave energy coming to us is generated by colder celestial bodies in the universe (force-centres with few or no satellites).

So, when you see a *yellow* towel here on earth, your eyes are detecting the EME radiated by the sun at a wavelength of ≈6.3E-07m diffracted by the molecules in your towel, but at an intensity that will not harm your eyes.

2.2   The Proton-Electron Pair

A proton-electron pair is a single proton with a single orbiting electron.

Apart from 'H⁺', all atoms are collections of proton-electron pairs.

The orbiting electron absorbs electro-magnetic energy from its surroundings and transforms it into kinetic energy. It then transfers this [kinetic] energy to its proton via their opposite electrical charges ($\pm e$).

The proton collects this energy as an operational electrical charge ($+e'$) via its magnetic charge (m_p). The proton transforms 'electron-kinetic' energy into electro-magnetic energy of the same magnitude.

The pair generates a constant magnetic field (due to its constant magnetic charge) that radiates from the positive (North) face of the orbital plane and around to the negative (South) face of the orbital plane (Fig 1) and which holds adjacent atoms together in the form of viscous matter.

The pair also generates a variable electrical field (due to a variable electrical charge) that works contrary to the magnetic field and acts as the carrier for electro-magnetic [energy] radiation (Fig 2). The electrical charge in a proton, which rises with temperature, also drives adjacent atoms apart in the form of gaseous matter (Coulomb's law).

This is why the density of matter alters little with temperature, but as a gas, adjacent atoms repel each other with a force proportional to their heat. If the magnetic field energy is greater than the electrical charge energy, the atoms will exist in a viscous form; the lower the temperature the greater the viscosity. If the electrical charge energy is greater, the atoms will exist as a gas; the higher the temperature, the greater the inter-atomic force (pressure; refer to Chapter 2.3.3).

Electro-magnetic energy is what we currently refer to as heat and light, and it can *only* be transmitted by proton-electron pairs. Neither a lone proton (H⁺) nor a lone electron has this capability.

The combined electro-magnetic energy from all the proton-electron pairs in an atom is its *quantity* of heat.

An atom's *temperature* is what we currently refer to as its highest electro-magnetic energy, which is radiated by the [two] proton-electron pairs whose electrons are orbiting in the innermost atomic shell.

Electro-magnetic energy rises and falls with the kinetic energy in the electron, which rises and falls with the energy in the radiation feeding it. In this way, [heat] energy in matter naturally stabilises with the energy in its environment. In outer-space, where there are no proton-electron pairs and negligible electro-magnetic energy, matter will naturally fall to ≈0 K

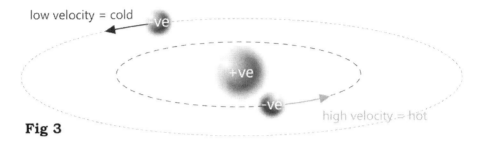

Fig 3

According to Newton's laws of orbital motion, an electron's orbital radius decreases with increasing kinetic energy (Fig 3). In other words, atomic structural strength increases with temperature.

2.3   The Atom

An atom is a collection of proton-electron pairs that were fused together under very high pressure inside the ultimate-body, which is the only mass capable of overcoming the coupling ratio. Lighter elements are created by the reduction of heavier elements through fission in stars.

Elements are created by pushing the proton(s) of one proton-electron pair (or element) inside the electron shell(s) of another. This process is called fusion. Fusion does *not* generate energy; it requires energy input to work.

Different elements have different numbers of proton-electron pairs, defined by their atomic number (Z); i.e. the number of protons inside a common group of electron shells. Its atomic number defines an element's character.

The similar [positive] electrical charges in protons prevent them from sitting together inside a common group of electron shells unless they are electrically isolated from each other; this is the job of the neutron.

The minimum number of neutrons in any nucleus is the same as the number of protons (refer to Chapter 4.4). I.e. if an element has an atomic number of 20, it can have no less than 20 neutrons. However, as proton-electron pairs can exist as deuterium (1 proton and 1 neutron) *and* tritium (1 proton and 2 neutrons) or any combination thereof, all elements will comprise between one and two neutrons per proton.

But there are limits to an atom's neutronic ratio (ψ): whilst 'ψ' must be greater than 1, if in practice it exceeds, or is even close to 1.5, the atom will frequently shed some of its neutrons.

As neutrons cannot exist outside an atom, when ejected they will divide into their component parts as alpha (proton) and/or beta (electron) particles. The emission of alpha-particles is called "radiation".

Ratios greater than 1.6 will cause an atom to break apart, into smaller atoms, also resulting in the emission of alpha-particles.

All atoms with an atomic number greater than 92 are unstable and will decay into smaller atoms relatively quickly. When this happens, it is usually accompanied by the simultaneous ejection of numerous neutrons.

The nesting of neutrons within a nucleus can affect an element's ability to eject them. If trapped within a large enough nucleus with plenty of neutrons to provide the necessary shielding, whilst a neutron will decay into its component parts, the atom will have acquired a new proton-electron pair that will emit its stored energy as electro-magnetic radiation, generating 'heat' and changing its identity. This is what occurs inside a nuclear reactor.

The amount of energy an atom can hold at any given time is shared between its proton-electron pairs according to their shell radii. The highest energy level is held and radiated by the innermost pair and the lowest is held and radiated by the outermost pair.

The sum of all electron energies in an atom define its heat energy. The maximum energy that can be held by all its electrons without the innermost proton-electron pairs transforming into neutrons is called an atom's heat capacity.

Electrons ejected from an atom will hold their linear (v) and angular (ω) velocities at the time of ejection in free-flight until affected by magnetism, and/or electro-magnetic energy. What we see in bubble chambers as post-impact spiral paths is simply the result of impacting electrons that can be visualised as spinning billiard balls obeying Newton's laws of motion.

And that is it; the atom really is that simple!

2.3.1 Electron Shells

Each electron shell contains up to two electrons, which are identical. An orbiting electron is held in its shell by a balancing act between its centrifugal force and the electrical attraction to its proton (opposite charges), just as Newton described it for an orbiting planet and its star.

Each orbiting electron is diametrically opposite to each other within its shell. All electrons are identical, except for their kinetic energy if they are not in the same shell.

Shell radii are defined by electron velocity, which is defined by electron kinetic energy, which is defined by surrounding heat (electro-magnetic energy). The highest electron energies are in the innermost shells, gradually reducing with radial distance from the nucleus. Spacing between each shell is identical as it is determined by the repulsion forces between adjacent electrons of identical charge. Principal shell radii are defined as follows:

The maximum *possible* shell radius (R_c) is achieved when an electron's velocity (v_c) falls to a level whereby it is expected to leave its orbit and continue in free-flight [3]

Planck's maximum shell radius ($R_o = a_o.(4\pi)^2$) is Planck's maximum orbital radius, the relevance of which is as yet unknown, but happens to be the innermost shell radius of Radon (the largest noble gas) at its gas transition temperature [1]

The minimum *possible* shell radius (R_n) is achieved when an electron's velocity (c) causes it to combine with its proton and create a neutron

Planck's mean radius (R_m), is defined by his modified constant (h' : refer to Chapter 5.4) and falls between R_o and R_n and defined as follows:

$R_m = 2.h'.\xi_v / m_e.c^2$
But was originally defined by Rydberg and can be equated thus:
$R_m = R_n / a_o.R_\infty$

And unites Planck's and Rydberg's constants thus:
$h' = \frac{1}{2}.m_e.c^2 . R_n / R_\infty.a_o.\xi_v$
given that $R_\infty = 1 / a_o.\xi_v$
$h' = \frac{1}{2}.m_e.c^2 . R_n \{J.m\}$

2.3.2   Nucleus

The nucleus of an atom contains protons and neutrons that are held together by magnetic field energy. It is probable that a proton and a neutron may touch (physically) in the nucleus as the neutron acts as a barrier to the positive charge between neighbouring protons.

Nucleic organisation is dependent upon the isolation of its proton's positive charges by attached neutrons.

This nucleic organisation is mirrored in an element's lattice structure. For example; what is currently referred to as a close-packed-hexagonal (HCP) lattice structure means that the protons within the nucleus of these elements are also arranged in the same formation.

The potential number of proton-electron pair orientations is limited if proton separation is to be ensured. This limitation defines the nucleic structure and the magnetic field energy arrangement between adjacent atoms. The magic number is '9' (refer to Chapter 3.3.2)

Because of the mirroring of atomic nuclei and elemental lattice structures, their mathematical relationships are also the same.

Regardless of the nucleic pattern, structural integrity generally reduces with increasing nucleic size (i.e. as the atomic number increases), making larger atoms generally (but not necessarily) more unstable due not only to the greater potential for proton-proton interaction but also the neutron-neutron interaction. This latter problem generates radioactivity; the greater the neutron-neutron interaction, the higher the radioactivity. Technetium appears to be a special case, in which $\psi = 1.3$ (refer to Chapter 3.3.2) but is especially susceptible to neutron decay. However, technetium is not a natural atom, it decays very quickly into a different element; it is highly radioactive.

2.3.3   The State of Matter

The nucleic lattice structure applies to, and affects, the opposing electrical and magnetic force configuration of all atoms. Therefore, the structural configuration of adjacent atoms in both viscous and gaseous forms is defined by its nucleic lattice structure (Refer to Chapter 3.3.2).

Matter exists either in viscous form (solid and liquid) or as a vapour (gas).

The electrical charge developed within a proton whilst hosting an orbiting electron (e') varies between 'e' at minimum temperature and ξ_v.e at maximum temperature; i.e. immediately before the two particles unite as a neutron. It is this charge (e') that repels adjacent atoms.

The magnetic field energy generated by the proton-electron pair holds adjacent atoms together. This field energy is constant because the magnetic charges generating it are constant (m_e & m_p).

These two competing energies are responsible for the state of matter:

1) If the electron in a proton-electron pair is orbiting slowly (low temperature), the electrical repulsion force between adjacent atoms will be less than the attraction force generated by the magnetic field energy. In this case, matter will exist in viscous form; the lower the electron's energy (temperature), the greater the viscosity of the substance.

2) If the electron in a proton-electron pair is orbiting quickly (high temperature), the electrical repulsion force between adjacent atoms will be greater than the magnetic attraction force. In this case, matter will exist as a gas; the greater the electron's energy (temperature), the greater the repulsion between adjacent atoms; or the higher the gas pressure.

This is the reason the density of viscous matter remains fairly constant, even with varying temperature whilst gas pressure rises and falls proportionally with temperature variation.

Given that the same forces (electrical and magnetic) are generating the attraction/repulsion of all adjacent atoms, one would expect the Periodic Table to reflect this, which it does (Fig 4).
However, 'Fig 4' also shows us:
1) how relative neutron-population (ψ) affects density ($\rho \propto \psi \propto Z$); and,
2) the effect of a constant magnetic charge on gas transition point

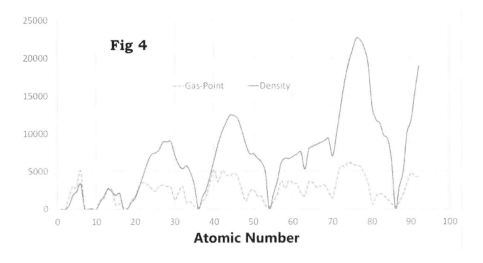

Fig 4

Atomic Number

Viscosity

Atomic density at low temperatures is considerably less than matter density. For example, the density of iron is 7870 kg/m³
whilst the density of its atom:
@ 273.15 K: ρ = 0.083888668 kg/m³
@ 12,412 K: ρ = 7870 kg/m³

At relatively low temperatures, this 'electron-clouding' allows lone protons (positive ions) to share spare electron charge capacity in neighbouring atoms, but diminishes with increasing energy. I.e. chemical bonding weakens with increasing temperature.

Viscosity is generated [in non-gaseous matter] by the competing [attractive] magnetic and [repulsive] electrical forces between adjacent atoms. The greater the difference between the magnetic fields and electrical charges, the greater the viscosity of matter.
Or as the temperature of the matter rises, its viscosity falls.

Gas

Dalton's law states that each gas in a mixture of different gases will be evenly distributed within its container independently of all the other gases in the mixture.

Partial pressure theory states that the total pressure of a gas mixture is the sum of the pressures of each individual gas.

These two theories (or statements) *appear* to conflict:

If Dalton's theory was strictly correct, the pressure of the gas mixture would be the maximum individual pressure, not their sum.

However:

The pressure in a gas, or mixture of gases, is generated by Coulomb's force law [repulsion] between all the protons in the gas mixture due to their electrical charges of similar polarity (+e').

All atoms (or molecules) with the same nucleic lattice structure will repel each other equally, generating an even distribution within their container: **Dalton's law**

All atoms (or molecules) with different nucleic lattice structures will repel each other, but their forces will not balance: **Partial pressure theory**

The reason partial pressure theory works is because all the atoms (or molecules) in a gas mixture repel all other gas atoms (or molecules) irrespective of their proton population, but only those with the same nucleic lattice structure will balance (distribute evenly).

This is why the total pressure of a mixture of gases can be established from the sum of all individual [balanced] gas pressures, *and* each individual gas may be treated as independently filling the container.

2.4   Isotope

Isotopes are atoms with the same atomic number (Z) but with varying atomic mass because of unequal proton-neutron pairing. Isotope is an alternative way of saying relative atomic mass (RAM).

An atom of iron, with 26 protons (Z=26) and 26 neutrons (N=26) is an isotope of 52. However, in nature, most iron atoms have more than 26 neutrons, each of which is allotted its own isotope, e.g. 57, 59, etc.

The following rules apply to isotopes:
1) H^+ can never be fused because they always repel each other
2) All proton-electron pairs within atoms are Deuterium or Tritium
3) Theoretically: $1 < \psi < 2$; Practically: $1 < \psi < 1.6$ (see below)

Despite the theoretical maximum value for ψ = **2**:

*If an atom achieves a 'ψ' value of greater than **1.5** it will readily and rapidly eject neutrons as alpha and beta-particles.*

*If an atom achieves a 'ψ' value of greater than **1.6** it will split into smaller atoms ejecting numerous alpha and beta-particles as it does so.*
***1.6** is the limiting number for atoms.*

Over time, atoms naturally try to achieve ψ = **1**, which is their most stable form. They eventually achieve this by ejecting surplus neutrons as alpha and beta-particles. The rate at which this occurs is referred to as the 'half-life' of the atom.

However, if all atoms in the universe had a neutronic ratio of 1, there would be no chemical reactions (including life).

2.5 Ion

Ions are atoms with the same atomic number (Z) but possess an electrical charge owing to unequal proton-electron pairing.

Positive ions (atoms that have lost electrons) possess a positive electrical charge. Negative ions (atoms with additional electrons) possess a negative electrical charge. Negative ions are far less common than positive ions.

Only a few atoms exist naturally as negative ions and they are all non-metals$_n$ except for two, which are semi-metals$_s$:

One additional electron (Group VIIA):
Fluorine (9_n), Chlorine (17_n), Bromine (35_n), Iodine (53_n)

Two additional electrons (Group VIA):
Oxygen (8_n), Sulphur (16_n), Selenium (34_n), Tellurium (52_s)

Four additional electrons (Group IVA):
Carbon (6_n), Silicon (14_s).

Any atom can become a positive ion simply by losing one or more of its electrons from impact with free electrons or a strong external positive electrical charge.

Negatively charged ions are a little more difficult to understand. Additional electrons need to be trapped by the positive charge in protons that do not exist in the nucleus: this shouldn't be possible. However, the nucleic structures of the above non-metal atoms probably have at least one exposed proton that is not protected by a neutron and this means that the additional electro-magnetic electrical charge generated in it is available to trap passing free electrons

2.6   Radioactive Fission

As explained in Chapter 2.5, the isotopic decay of atoms is the disintegration of neutrons into their component parts; a proton and an electron. This process is called *fissionable nuclear radioactivity* and simultaneously releases the energy that was stored in the neutron (refer to Chapter 2.7.3).

Alpha and beta particles (protons and electrons) are normally ejected in pairs because neutrons are always created in pairs. As there are always two electrons in the innermost electron shell, both of them will possess the same kinetic energy at the same time. Therefore, when one innermost orbiting electron achieves velocity 'c', so will its neighbour. This is why an alpha particle is commonly referred to as an helium atom (two protons).

This energy is released in the form of kinetic energy in high-speed protons (and usually electrons) or if prevented from ejection, as electro-magnetic energy (heat) due to the re-creation of additional proton-electron pair(s) altering an element's atomic number (Z).

The kinetic energy in an ejected proton is sufficient to detach neighbouring neutrons; a chain-reaction of alpha particles.

The energy released by a restored proton-electron pair that remains within its atom will be in the form of electro-magnetic radiation (heat).

This process occurs naturally in all atoms with a neutronic ratio (ψ) greater than 1. Usually, the higher the neutronic ratio the more frequently it occurs. What we call a radioactive element is one that represents a danger to life as we know it; i.e. $\psi > 1.5$ and will continue to eject neutrons unhindered until all the atoms in the matter have achieved $\psi = 1$

When this process is enforced artificially, using a radioactive element's *critical mass*, neutronic energy release can be significantly accelerated.

As critical mass is approached *slowly*, many of the neutrons will revert to their original proton-electron pairs, resulting in a consequent increase in heat within the matter. Or, if it is reached quickly enough, the neutrons will be ejected as protons (and electrons) breaking the matter apart. This is called an explosive chain reaction and is what occurs in an atom bomb.

Our knowledge of radioactivity today only allows us to release such energy with significant risk. However, processed correctly, it can be released safely in a controlled manner, allowing us to generate fission-energy from non-radioactive matter and even to recycle our existing nuclear waste.

Radioactivity can, and should be regarded as a friend, not an enemy, especially as free energy is *only* available from fission (not fusion); you just need to know how to harness it (refer to Chapter 2.8).

2.7   Atomic Particles

There are only two atomic particles; the proton and the electron, and there are more than 2.8E+75 of each of them in the universe. All protons are identical and all electrons are identical. There is no uniqueness or uncertainly regarding their properties or their behaviour. They are all entirely predictable.

Protons and electrons have the same ultimate *mass* density (refer to Appendix A3)

2.7.1 The Proton

The proton is a static particle that possesses a constant magnetic charge, which is 'ξ_m' times more massive that the electron. We currently refer to this magnetic charge as a proton's *mass*.

As a hydrogen atom (H$^+$), a lone proton possesses the same electrical charge (e) as an electron.

When paired with an orbiting electron, however, its additional magnetic charge (*mass*) enables it to collect additional electrical charge energy from its orbiting electron that the pair use to generate electro-magnetic energy.

2.7.2   The Electron

The electron is a dynamic particle that possesses a constant magnetic charge, which is 'ξ_m' times less *massive* that the proton. We currently refer to this magnetic charge as an electron's *mass*.

An electron always possesses the same electrical charge (e).

When paired with a proton, an electron will collect electro-magnetic energy radiated by other proton-electron pairs, which it transforms into kinetic energy [1].

2.7.3   The Neutron

The neutron is a proton and an electron held together by magnetic energy and possesses 1.637856E-13 Joules of energy. It is created within an atom when the innermost proton-electron pairs achieve a temperature that unites both particles, and destroyed when ejected from the atom.

A neutron's component parts are the alpha (proton) and beta (electron) particles ejected during radioactive decay.

It has three primary functions;

1) It allows atoms to exist, by shielding proton charges and thereby permitting protons to sit together inside the same electron shells, and;

2) It provides most of the heat and light emitted by the stars (fission), and;

3) It provides the energy for the next *Big Bang*

The neutron is, of course the very essence of universal existence, making the neutron a far more important particle than initially appears. In fact, without the neutron, there would be no elements and no *Big Bang* and therefore no universe as we know it!

2.8　　The Early Atom

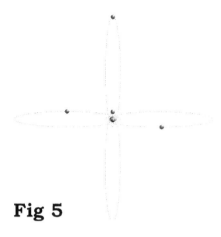

For almost a hundred years, we have been taught that an atom comprises a nucleus of protons and neutrons with electrons orbiting the nucleus in elongated elliptical shells, as proposed by Johannes Rydberg.

His inner shells contained the least number of electrons and the outer shells contain the most. Every electron in the atom must be unique, s, p, d, f, l, m, spin, etc.

Fig 5

It is an enigma (to me), how this level of complexity ever became proposed and/or accepted, because it simply doesn't work. For example,

1) If Rydberg's shell system is fully analysed, it produces shells of eccentricity equal to 'one'; i.e. a straight line!

2) Elliptical orbits mean variable velocities, which would mean fluctuating electro-magnetic radiation (e.g. colours), which does not occur in reality

3) Electrons in elliptical orbits cannot be equally spaced, and therefore balanced between their similar static electrical charges. This is only possible with circular orbits.

4) Elliptical orbits only work where magnetic potential energy is greater than repulsive electrical energy, which is not the case in an atom.

5) Electrons are not driven by potential energy, as planets are.

6) If the calculated eccentricities are correct, according to elliptical theory; $(b = a.\sqrt{[1 - e^2]})$, the apogee radii reduce with shell number ultimately becoming smaller than the perigee radius of shell-1
(e.g. Shell 6: 2.58899E-11 m)

7) Shell radii calculated according to this theory cannot be reconciled with generally accepted atomic radii for most elements

8) Photographs of atoms appear to show their structure to be spherical, not elliptical

It doesn't matter how the calculations are carried out Rydberg's system cannot be made to work.

Until Quantum theory diverted our attention from reality, we were taught that the atom looked something like the image in Fig 5. The need for an alternative structure came about because Johannes Rydberg's model generated an electron shell eccentricity of 1 (a straight line), which could not be explained or resolved.

Unfortunately, the scientific community unanimously decided that *Quantum Theory* must provide the answer. String theory, sub-atomic particles and the uncertainty principle were subsequently invented to justify it.

This model gave us the fuzzy, non-uniform, statistical, unpredictable, complex shape we frequently see in textbooks today.

In fact, not only are these theories unnecessary, if all matter in the universe actually did obey them, the universe as we know it could not exist. The atom would not work because it could not emit electro-magnetic energy.

After discovering that Newton's laws of orbital motion apply to *all* orbital systems, I decided to see if his theories could be applied to the atom and discovered very quickly that they can.

In fact, the only atom that genuinely works is the Newton-Coulomb atom, and it doesn't need sub-atomic particles to make it work.

So, the atom actually works according to the laws of Gilbert, Coulomb, Faraday, Maxwell, Lorentz and Newton, which could (and should) have been discovered before relativity and quantum theory stalled scientific progress at the beginning of the twentieth century.

3 Calculation Procedures

This section is a compilation of the mathematical formulas supporting the narrative, including how to use them. It has been written to simplify their use.

3.1 Energy

The energy held within a neutron is that which was generated within the proton-electron pair at the time of their union and is a combination of:

The kinetic energy in an orbiting electron

The potential energy between a proton and its orbiting electron

The spin energies in the proton and its orbiting electron

This energy; $E = |KE| + |PE| + |SE|$ is trapped within a neutron when its orbiting electron achieves '*light-speed*' (c) and released again when the neutron is ejected from its atomic host due to radioactivity; whether naturally or artificially induced.

3.1.1   Potential Energy

Its linear mathematical relationship is: $PE = m.g.R = \frac{1}{2}.m.v^2$

In circular orbits, such as atoms, potential energy between particles is *always* twice the kinetic energy in the orbiting satellite, so their mathematical relationships is:

$PE = CE = 2 . \frac{1}{2}.m.v^2 = m.v^2$

At the *speed of light* this becomes $PE = m.c^2$, which is the true meaning of Henri Poincaré's formula and applies to *orbiting* electrons

3.1.2 Kinetic Energy

Its general mathematical relationship is: $KE = \frac{1}{2}.m.v^2$

The kinetic energy of a satellite orbiting in a circular path is exactly half the satellite's potential energy: $KE = \frac{1}{2}.PE = \frac{1}{2}.m.v^2$

3.1.3 Electro-Magnetic Energy

Electro-magnetic radiation always travels at the same **velocity**, which we currently refer to as the *speed of light*: c = 299792459 m/s, but is of course, the same speed as *all* electro-magnetic energy.

Knowing this, and that its energy is exactly the same as the kinetic energy in the electron that transferred it, we can calculate its other properties:

Wavelength (λ) is defined by the orbital velocity (v) of the electron transferring it: $\lambda = 2\pi R.c/v = c/f$ {m}
Where: R is the electron orbital (shell) radius

Johannes Rydberg gave us a relationship between wavelength and shell number (n) that actually works:
If the electron temperature (T_1) in shell number 1 (n=1) is known, the electron temperature (T_n) in any shell may be calculated as follows:
$T_n = T_1/n$
For example; the electron temperature at Rydberg's orbital radius (a_o) ...
$T = X_R/a_o = 33192.4000063507$ K
Rydberg's constants may be used to calculate the electro-magnetic wavelength and energy generated by an electron in any shell as follows:
$E_n = R_y . T_n/T$
$\lambda_n = \frac{1}{2} . (T/T_n)^{1.5} / R_\infty$
Where: X_R (heat transfer coefficient) = 1.75646616508036E-06 K.m

Frequency (f) is defined by the orbital period of the electron transferring it: $f = v/2\pi R = c/\lambda$ {Hz}

Amplitude (A) is equal to the orbital radius of the electron transferring it: A = R {m}

Energy (E) may be determined using the modified version of Planck's constant or the orbital velocity (v) of the electron transferring it:
$E = h'/A = \frac{1}{2}.m_e.v_e^2$ {J}

A proton develops an operational **Charge** (e') whilst hosting an orbiting electron that maximises at: $e_{max} = e.\xi_v$
$e' = e . v/v_o$ {C}
This charge is used to generate electro-magnetic radiation it emits and varies with the kinetic energy of the orbiting electron responsible for it.

3.2   Proton-Electron Pair

A proton-electron pair is an electron orbiting a proton that obeys Newton's laws of orbital motion and Coulomb's force law, the difference between them is the coupling ratio (φ).

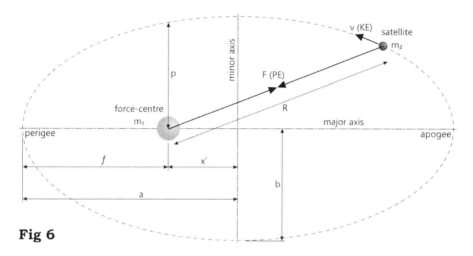

Fig 6

Newton's laws of orbital motion are based upon non-circular elliptical orbits (Fig 6), but because the electron provides its own kinetic energy, its orbit will be circular (Fig 7). A circular orbit follows the same mathematical principles as an elliptical orbit but its shape is much simpler, eliminating the need to calculate many of its properties.

I.e.

R: radial distance between the satellite and force-centre centres of mass

a = **b** = **p** = f = **R**: orbital radius

e: eccentricity of the ellipse = 0

x': distance from focus to ellipse centre = 0

Therefore we can ignore; a, b, p, f, e & x'

3.2.1   Input Data

Before calculating the properties of an orbit, we must first identify the input data relating to atomic particles; i.e. the information required to start the calculation. The input information is usually given as follows because it is the easiest to define.

The Variables:

m_p: proton mass = 1.67262163783E-27 kg

m_e: electron mass = 9.1093897E-31 kg

T: temperature; magnitude of electromagnetic energy (heat) absorbed by the electron

The Constants:

X: Heat constant; 6.9353271647894E-09 K.s^2/m^2

X_R: Heat constant; 1.75646616508036E-06 K.m

φ: coupling ratio = 4.40742111792335E-40

G: Newton's gravitational constant; 6.67359232004332E-11 m^3/kg/s^2 (refer to Chapter 6.11.2)

3.2.2   Orbital Shape

The relative electrical charge:
$RAC = e/m_e$

The velocity of an orbiting electron:
$v = \sqrt{[T/X]}$

The innermost orbital radius:
$R_1 = X_R/T$
Each subsequent orbital shell radius
may be calculated thus:
add this radius (R_1) to each
subsequent shell radius (R_2 to R_n)

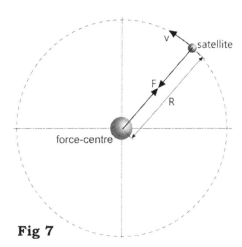

Fig 7

The following relationship is also true
for electron shell radii:
$R = G.m_p / \varphi.v^2$

The potential acceleration in the proton-electron pair is:
$g = v^2/R$

An electron's orbital period:
$t = 2.\pi.R/v$

The orbital area and the orbital path length:
$A = \pi.R^2$
$L = 2.\pi.R$

An important fact to remember about the constant of proportionality (K) is
that it remains constant for all electrons irrespective of shell number
(radius) or velocity (temperature).
It may be determined as follows:
$K = t^2/a^3 = 0.15587874533403$ {s^2/m^3}
where 't' is the orbital period and 'a' is half the orbit's major axis.

The above constitutes everything you need to determine the size of every
electron shell in an atom. Each shell can hold up to two identical
electrons, which is confirmed by 'ℓ' above being the same as 'R'.
m_p can now be confirmed using 'K' and 'G'.

3.2.3 Body Mass

The proton and electron masses are included in the input data for atomic shell calculations. However, it is handy to know that m_p can be confirmed as follows.

Force-centre mass may be calculated from the first orbital analysis thus:
$m_p = \varphi.(2.\pi)^2 \, / \, G.K$

or you may use Isaac Newton's famous formula if you prefer:
$m_p = \varphi.g.R^2 \, / \, G$

This is the only time you need to apply Newton's gravitational constant (G); i.e. m_p (and 'K') remains the same for all other electron orbits, irrespective of electron velocity and shell number.

The coupling ratio (φ) is needed in the above formulas because it is the electrical charge that is holding the electron to the proton, not magnetism.

We already know 'm_e' from the input data.

3.2.4 Electron Performance

Electron performance is constant throughout its orbit.

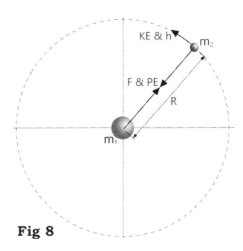

F: attractive force between the proton and the electron (-ve)

PE: attractive energy between the proton and the electron (-ve)

KE: kinetic energy in the electron (+ve)

E: total energy in the orbital system

Fig 8

p: electron momentum, which varies with orbital velocity

h: Newton's constant of motion (electron momentum without the mass component)

Potential [attractive] force:
$F = -g.m_e$

Centrifugal force in the orbiting electron:
$F_c = m_e.v^2/R$

Kinetic energy in the electron varies with orbital velocity:
$KE = \frac{1}{2}.m_e.v^2 (= \frac{1}{2}.PE)$

Potential energy between the electron and proton is twice the kinetic energy in the electron:
$PE = -2.KE (= F.R)$

Total energy is the sum of the two energies above:
$E = PE + KE$

Electron momentum:
$p = m_e.v$

Newton's constant of motion:
$h = R.\sqrt{[-PE / m_e]}$

3.3   The Atom

An element's natural ejection of neutrons is dependent upon two things:

1) its neutronic ratio (ψ)

and

2) its nucleic structure

Neutrons that are trapped inside a nucleus will often find it difficult to eject. In such a case, the *'ejected'* neutron will remain within the nucleus but will revert to its proton-electron pair. The atom concerned will then become a different element with an additional atomic number. For each neutron *'ejected'* in this manner, Z = Z+1

When this happens, -4.1E-14 J of [heat] energy will be released for each *'ejected'* neutron.

In the event a neutron is *actually* ejected, as a proton (alpha) and an electron (beta), they will both achieve an ejection velocity of:

$v = \sqrt{[\ 2.E\ /\ (m_p + m_e)\ ]}$ ($\approx$7E+06 m/s)

where $(m_p + m_e)$ = the mass of the electron and the proton

When the ejected proton impacts a neutron in another atom, it cannot impact another proton because of the similar [positive]) charge polarities, the character of impacted atom will alter (Z+1), which can cause natural molecules to fail.

The relationship between temperature (T), velocity (v) and orbital radius (R) may be defined as follows:

$v^2 = T/X$ and $R = X_R/T$

where: X & X_R are heat transfer coefficients (refer to Chapter 6.6)

It is the electro-static *potential* energy that holds an orbiting electron to its proton. This energy is *always* twice the kinetic energy in the electron:

$KE = \frac{1}{2}.m_e.v^2$

$PE = 2.KE = 2.\frac{1}{2}.m_e.v^2 = m_e.v^2$

At the time of ejection from its proton-electron pair:

Angular velocity in an electron is: $\omega = 2\pi$ / orbital period
(at the time of ejection)

The linear velocity of an electron is: $v = \sqrt{[2.KE / m]}$
(at the time of ejection)

The following calculation procedure provides the formulas required to convert atomic energy to electro-magnetic energy, electricity and the specific heat capacity for any atom.

There are a few important things to remember when carrying out these calculations:

1) The temperature we feel (and measure) is that emitted by the proton-electron pair(s) with its electron(s) in the inner-most electron shell (T_1). It is this temperature (only) that is used to define the [measured] specific heat capacity of an atom.

2) The specific heat capacity of an atom includes the kinetic energy of all of its electrons. The kinetic energy of the electron(s) in each shell will vary with its orbital radius.

3) All electrons possess the same electrical charge, which means that the spacing between each shell will be equal.

4) You start the calculation procedure by selecting the temperature of the atom; i.e. the temperature that you would feel or measure (T_1)

5) 'n' refers to the electron shell number; 1 to 46 for atomic numbers (Z) 1 to 92 (note: each shell can hold up to two electrons)

Input Data:

measured temperature: T_1

Atomic:

orbital radius of shell 1: $R_1 = X_R / T_1$
Properties in Shell 'n':
orbital radius: $R_n = R_1.n$
electron temperature: $T_n = X_R / R_n$
electron velocity: $v_n = \sqrt{[T_n/X]}$
kinetic energy of electron: $KE_n = \frac{1}{2}.m_e.v_n^2$
potential energy: $PE_n = -2.KE_n$

Electro-Magnetic:

properties of electro-magnetism radiated by [proton-]electron in shell 'n':
frequency: $f_n = v_n / 2\pi R_n$
wavelength: $\lambda_n = c/f_n$
amplitude: $A_n = R_n$
energy: $E_n = KE_n$
charge: $Q = e'$

Electrical:

current: $A_n = Q.f_n$
voltage: $V_n = E_n/Q$
resistance: $\Omega_n = V_n/A_n$
temperature: $T_n = 2.E_n / k_B.Y$ (will be same as T_n above)

Specific Heat Capacity (SHC):

$SHC = {}_1\Sigma^n KE / T_1.m.Y$ {J / kg.K}

Where:
m = atomic mass

3.3.1 Electron Shells

The innermost orbital radius is calculated thus; $R_1 = X_R/T$
then add this radius (R_1) to each subsequent shell radius (R_2 to R_n)

The electron temperature in subsequent shells can then be determined using the same formula: $T = X_R/R$

The orbital velocity of each electron is determined thus: $v = \sqrt{[T/X]}$

Their kinetic energies are: $KE = \frac{1}{2}.m.v^2$
Their potential energies are: $PE = -2.KE$

Electro-magnetic energy is equal to KE.

You use the formulas provided in Chapter 3.1.3 to find the properties of the electro-magnetic energy generated by each proton-electron pair.

We can calculate electron shell number (n) from an electron's kinetic energy thus:

$n = T_1/T_n = (v_1/v_n)^2$

Where; 'v' is the velocity of the electron

3.3.2　Nucleus

Atomic lattice structure and density are defined thus; $\zeta = \sqrt[3]{[\, m.n/\rho \,]} / R$
The resultant factors are listed below for Hydrogen to Uranium[4]:

Hydrogen	(1)	4.25	Silver	(47)	5.625
Helium	**(2)**	**1.5**	Cadmium	(48)	3.875
Lithium	(3)	19.5	Indium	(49)	6.125
Beryllium	(4)	16.25	Tin	(50)	6.875
Boron	(5)	17.5	Antimony	(51)	5.625
Carbon	(6)	18	Tellurium	(52)	4.75
Nitrogen	(7)	5.5	Iodine	(53)	3
Oxygen	(8)	5	**Xenon**	**(54)**	**2**
Fluorine	(9)	4.625	Caesium	(55)	6
Neon	**(10)**	**1.625**	Barium	(56)	6.875
Sodium	(11)	11.125	Lanthanum	(57)	8
Magnesium	(12)	10	Cerium	(58)	6.625
Aluminium	(13)	11.875	Praseodymium	(59)	7.875
Silicon	(14)	14	Neodymium	(60)	7.25
Phosphorus	(15)	6	Promethium	(61)	6.375
Sulphur	(16)	6.5	Samarium	(62)	5.5
Chlorine	(17)	6.5	Europium	(63)	5.875
Argon	**(18)**	**2.375**	Gadolinium	(64)	7
Potassium	(19)	10	Terbium	(65)	6.875
Calcium	(20)	10.75	Dysprosium	(66)	6.125
Scandium	(21)	12	Holmium	(67)	6.125
Titanium	(22)	10.375	Erbium	(68)	6.25
Vanadium	(23)	9.5	Thulium	(69)	5.25
Chromium	(24)	8	Ytterbium	(70)	4.625
Manganese	(25)	6.875	Lutetium	(71)	6.5
Iron	(26)	7.875	Hafnium	(72)	6.75
Cobalt	(27)	7.5	Tantalum	(73)	6.75
Nickel	(28)	7.5	Tungsten	(74)	6.5
Copper	(29)	7	Rhenium	(75)	6.375
Zinc	(30)	4.875	Osmium	(76)	5.875
Gallium	(31)	7.375	Iridium	(77)	5.5
Germanium	(32)	8.5	Platinum	(78)	5.25
Arsenic	(33)	4.375	Gold	(79)	4.75
Selenium	(34)	4.875	Mercury	(80)	2.375
Bromine	(35)	5.375	Thallium	(81)	4.125
Krypton	**(36)**	**2.25**	Lead	(82)	4.5
Rubidium	(37)	7	Bismuth	(83)	4.5
Strontium	(38)	7.625	Polonium	(84)	3.75
Yttrium	(39)	9.5	Astatine	(85)	3
Zirconium	(40)	9.5	**Radon**	**(86)**	**2**
Niobium	(41)	8.875	Francium	(87)	4.75
Molybdenum	(42)	8.25	Radium	(88)	5.625
Technetium	(43)	7.625	Actinium	(89)	6
Ruthenium	(44)	7.25	Thorium	(90)	7
Rhodium	(45)	6.875	Protactinium	(91)	5.75
Palladium	(46)	6.25	Uranium	(92)	5.5

As the above list shows; the lowest values for ζ' occur at **the noble gases**.
Refer to Chapter 3.3.3; Lattice Structure for an explanation of these
factors.

The factors that define an atomic lattice structure ($\Gamma = 9.[\ \psi-1\ ]$) is listed below, from Hydrogen to Uranium[4]:

Hydrogen	(1)	0.07146	Silver	(47)	2.655612766
Helium	**(2)**	**0.011709**	Cadmium	(48)	3.0770625
Lithium	(3)	2.823	Indium	(49)	3.089020408
Beryllium	(4)	2.2774095	Tin	(50)	3.3642
Boron	(5)	1.45998	Antimony	(51)	3.485294118
Carbon	(6)	0.01605	Tellurium	(52)	4.084615385
Nitrogen	(7)	0.008614286	Iodine	(53)	3.54981566
Oxygen	(8)	0.0001125	**Xenon**	**(54)**	**3.984**
Fluorine	**(9)**	**0.998403**	Caesium	(55)	3.748015636
Neon	(10)	0.16173	Barium	(56)	4.070410714
Sodium	(11)	0.809811818	Lanthanum	(57)	3.932447368
Magnesium	(12)	0.22875	Cerium	(58)	3.742137931
Aluminium	(13)	0.679526308	Praseodymium	(59)	3.494387288
Silicon	(14)	0.054964286	Neodymium	(60)	3.636
Phosphorus	(15)	0.5842566	Promethium	(61)	3.393442623
Sulphur	(16)	0.0365625	Samarium	(62)	3.826451613
Chlorine	(17)	0.769235294	Europium	(63)	3.709142857
Argon	**(18)**	**1.974**	Gadolinium	(64)	4.11328125
Potassium	(19)	0.520247368	Terbium	(65)	4.005047077
Calcium	(20)	0.0351	Dysprosium	(66)	4.159090909
Scandium	(21)	1.266818571	Holmium	(67)	4.154819104
Titanium	(22)	1.581954545	Erbium	(68)	4.137220588
Vanadium	(23)	1.933630435	Thulium	(69)	4.034896957
Chromium	(24)	1.4985375	Ytterbium	(70)	4.248
Manganese	(25)	1.77769764	Lutetium	(71)	4.178915493
Iron	(26)	1.331653846	Hafnium	(72)	4.31125
Cobalt	(27)	1.6444	Tantalum	(73)	4.308645205
Nickel	(28)	0.865735714	Tungsten	(74)	4.360135135
Copper	(29)	1.721172414	Rhenium	(75)	4.34484
Zinc	(30)	1.6155	Osmium	(76)	4.527236842
Gallium	(31)	2.24216129	Iridium	(77)	4.466922078
Germanium	(32)	2.4215625	Platinum	(78)	4.509
Arsenic	(33)	2.433163636	Gold	(79)	4.439227215
Selenium	(34)	2.901176471	Mercury	(80)	4.566375
Bromine	(35)	2.546742857	Thallium	(81)	4.709255556
Krypton	**(36)**	**2.9495**	Lead	(82)	4.741463415
Rubidium	(37)	2.789464865	Bismuth	(83)	4.660523133
Strontium	(38)	2.752105263	Polonium	(84)	4.390928571
Yttrium	(39)	2.516734615	Astatine	(85)	4.233917647
Zirconium	(40)	2.5254	**Radon**	**(86)**	**5.024406977**
Niobium	(41)	2.394083415	Francium	(87)	5.071034483
Molybdenum	(42)	2.558571429	Radium	(88)	5.116193182
Technetium	(43)	2.701297674	Actinium	(89)	4.95788764
Ruthenium	(44)	2.673409091	Thorium	(90)	5.20381
Rhodium	(45)	2.5811	Protactinium	(91)	5.142857143
Palladium	(46)	2.821304348	Uranium	(92)	5.285436848

As the above list shows; 'Γ = 0 to 5' occur at **the noble gases** (except Neon).

3.3.3   The State of Matter

Atoms exist either as viscous matter (not solid-fluid-liquid) or as a gas.

In both viscous and gaseous forms, adjacent atoms are repelled by electrical force (F_E) and attracted by magnetic force (F_M).

In viscous matter; the magnetic force between adjacent atoms is greater than the electrical force. In gases; the electrical force is greater. The transition-point for all matter occurs when the magnetic and electrical forces are equal; $F_M = F_E$

The prominence of F_M & F_E will determine whether or not atoms exist in viscous or gaseous form.

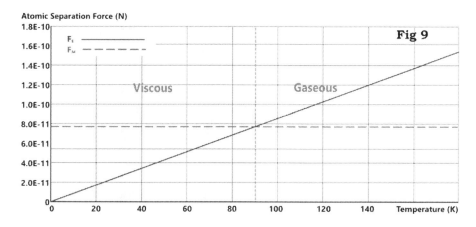

Fig 9 shows this transition from viscosity to gas for the oxygen molecule.

Notes:

Atomic Mass (m_A): For all following calculations, atomic mass is the mass of an atom, and may be calculated thus:
$m_A = m_n.RAM$

Molecular Mass (m_M): For all following calculations, molecular mass is the atomic mass multiplied by the number (N) that exist in the molecule:
$m_M = m_A.N$

m_n is the mass of a neutron

3.3.3.1 Lattice Structure (ζ)

An element's natural lattice (or crystal) structure is defined by its nucleus, which regulates the way adjacent *same-element* atoms interact. We currently refer to these structures as; Face Centre Cubic, Body Centre Cubic, Close-Packed Hexagonal, Tetrahedron, etc. The electrical repulsion force (**F$_E$**) between the protons in adjacent atoms is responsible for this structure.

Lattice structures apply to all *same-element* atoms in both viscous and gaseous forms. Dissimilar electrical repulsion patterns in nuclei that do not contain the same number of protons are responsible for preventing natural crystals occurring in dissimilar atoms (elements) and partial pressure theory (refer to Chapter 3.3.3.4).

A special variable 'ζ' defines the lattice structure of a collection of *same-element* atoms.

Radial separation between each adjacent atom in both viscous and gaseous forms is defined thus:

$R_L = \sqrt{[m_A / T_g.N.A]}$

Where 'A' is a constant associated with the lattice (crystal) structure and looks like this; $A = m_p / R_n.Y.T_n$ {kg / K.m^2}

$\rho = m_M / [ζ.R_L]^3$

$ζ = \sqrt[3]{[m_M/\rho]} / R_L$ {no units}

The lattice values (ζ) for each element are listed in Chapter 3.3.2

3.3.3.2 Inter-Atomic Forces

Isaac Newton's constant of motion (h) applies to both an orbiting electron:

$h_e = c.R_n$

and also to a stationary proton thus:

$h_p = c.R_n / \sqrt{\xi_m} = e.\sqrt{[\mu'.G/\varphi]}$ {m^2/s}

and may be applied to magnetic inter-atomic force thus:

$\mathbf{F_M} = h_p{}^2.m_A / R^3$

Electrical inter-atomic force may be calculated thus:

$\mathbf{F_E} = k_B.\underset{.}{T}.N/R = PE_1.N / R.Y$

Gas-transition (T_g) occurs when both of these forces are equal:

$h_p{}^2.m_A/R^3 = k_B.\underset{.}{T}.N/R$

Inter-atomic separation [distance] for any viscous matter is therefore:

$\mathbf{R_v} = \sqrt{[m_A / N.T_g.A]}$

Where 'A' is a constant associated with the lattice (crystal) structure and looks like this; $A = m_p / R_n.Y.T_n$ {$kg / K.m^2$}

The gravitational acceleration between adjacent atoms in a crystal, may be calculated using Newton's gravitational constant as follows:
$\mathbf{g} = G/\varphi . m_A/R^2$

or

$\mathbf{g} = \mathbf{F_M}.R / \mu'.e^2$

3.3.3.3 Viscous Matter

Viscous matter is that in which the magnetic force between adjacent atoms is greater than the electrical force. Whilst small changes in radial separation will occur when viscous matter alters lattice configuration, atomic radial separation, and therefore matter density, remains constant because magnetic force does not alter with temperature (Fig 9).

Identical electrical repulsion patterns in atoms with the same number of protons are responsible for restricting natural crystallisation to *same-element* atoms.

Radial separation (**R**) between adjacent atoms at all temperatures below 'T_g' is therefore constant, notwithstanding the aforementioned exceptions.

Electrical inter-atomic force may be calculated for any temperature thus:

$$\mathbf{F_E} = k_B.T.N \ / \ R = PE_1.N \ / \ R.Y$$

Magnetic inter-atomic force is dominant in all viscous matter and may be calculated thus (irrespective of temperature):

$$\mathbf{F_M} = h_p{}^2.m_A \ / \ R^3$$

The density of viscous matter may be calculated:

$$\mathbf{\rho} = m_M \ / \ (\zeta.R)^3$$

Where ζ defines the lattice arrangement (refer to Chapter 3.3.2)

3.3.3.4 Gaseous Matter

Identical electrical repulsion patterns in atoms with the same number of protons are responsible for the partial pressure phenomenon. I.e. just as only *same-element* atoms crystallise in viscous form; whilst all atoms repel each other, only the repulsion forces between *same-element* atoms will balance; generating a constant and equal pressure between *same-element* atoms throughout a mixture of different gases.

Adjacent atoms exist in gaseous form when the electron kinetic energy (temperature - heat) are such that electrical inter-atomic forces are greater than magnetic inter-atomic forces.

Electrical inter-atomic force may be calculated for any temperature thus:

$\mathbf{F_E} = k_B \cdot T.N^2/R$ $(= p.A = T.k_B.\rho.A / m_M$ **# see below**)

The density of gaseous matter may be calculated thus:

$\rho = m_M / (\zeta.R)^3$

Today, we calculate the pressure (p) of a gas thus:
$p = n.R_i.T / V$
where: 'n' is the number of moles in the gas, 'R_i' is the ideal gas constant, 'T' is its temperature and 'V' its volume.

But it can also be calculated thus:
$p = \rho.PE_1 / m_M.Y = T.m_e.\rho / X.Y.m_M = k_B.T.\rho / m_M$ **#**

where: 'ρ' is the gas density, 'PE_1' is the potential energy between the proton and the electron in shell-1 and 'm_M' is the molecular mass:
$PE = m_e.v_e^2$
$v_e = \sqrt{[T/X]}$

which provides *exactly* the same result as the PVRT calculation method but is much simpler because there is no need to play with moles.

Because this latter calculation method replaces 'PVRT' altogether, along with the need for Boltzmann's constant, Avogadro's number, gas temperature and the ideal gas constant with potential energy, the model described here must be considered correct.

3.3.4 Planck Properties

An electron absorbs electro-magnetic energy and converts it into velocity. This means that at the '*speed-of-light*' its kinetic energy reaches:
$KE = \frac{1}{2}.m_e.c^2$
and the coincident *potential* energy between the proton and its electron is:
$PE = m_e.c^2$

There are three key energy conditions for the proton-electron pair according to Planck, each of which relate to electron velocities, shell radii and associated temperatures. These energies are listed below:

Neutronic: $KE_n = \frac{1}{2}.m_e.c^2 = 4.0935556113127E-14$ J

Mean: $KE_m = \frac{1}{2}.m_e.v_m^2 = 2.3771466644364E-17$ J

Minimum: $KE_o = \frac{1}{2}.m_e.v_o^2 = 1.3804200555196E-20$ J

There is also a fourth energy level that appears to be when an orbiting electron may leave its proton partner and continue in free-flight. This condition is referred to as the '*cold*' energy level:

Cold: $KE_c = \frac{1}{2}.m_e.v_c^2 = 8.0161630672150E-24$ J

$\delta = KE_n / KE_c = 5.1066271693683E+09$

The potential energy between a proton and its orbiting electron is twice the kinetic energy of the electron.

The energy stored within each neutron is therefore: $-KE_n = PE_n + KE_n$

The orbiting electron defines the properties of the electro-magnetic energy emitted by the proton-electron pair (refer to Chapter 3.1.3)

Note: subscript 'n' in the following sub-chapters refers to the neutronic condition, not the shell number.

3.3.4.1 Planck Electron Velocities

Neutronic: c = 299792459 m/s

Mean: $v_m = \sqrt{[T_m/X]}$ = 7224342.80705 m/s

Minimum: $v_o = \sqrt{[T_o/X]}$ = 174090.866621 m/s

Cold: $v_c = \sqrt{[T_c/X]}$ = 4195.2092599072 m/s

This means that an electron is unlikely to exceed (or even achieve) the *'speed of light'* in free-flight unless it is provided with artificial energy greater than $\frac{1}{2}.m.c^2$ or knocked from its orbit, and most electrons in free-flight will be travelling at little more than 4000m/s

3.3.4.2 Planck Shell Radii

Neutronic: $R_n = X_R/T_n = 2.81793795383896E\text{-}15$ m

Minimum: $R_m = X_R/T_m = 4.85261843362263E\text{-}12$ m

Mean: $R_o = X_R/T_o = 8.35643156381571E\text{-}09$ m

Cold: $R_c = X_R/T_c = 1.43901585166681E\text{-}05$ m

The minimum possible or orbital radius cannot be less than; R_n

3.3.4.3 Planck Temperatures

Neutronic: $T_n = X.c^2 = 623316124.71718$ K

Mean: $T_m = X.v_m^2 = 361962.55467156$ K

Minimum: $T_o = X.v_o^2 = 210.19332853584$ K

Cold: $T_c = X.v_c^2 = 0.122060237421696$ K

The maximum possible natural temperature cannot exceed; T_n

3.4   Isotope

An element's natural ejection of neutrons is dependent upon two things:

1) its neutronic ratio (ψ)

and

2) its nucleic structure

Neutrons that are trapped inside a nucleus will often find it difficult to eject. In such a case, the '*ejected*' neutron will remain within the nucleus but will revert to its proton-electron pair. The atom concerned will then become a different element with an additional atomic number. For each neutron '*ejected*' in this manner, Z = Z+1

When this happens, -4.1E-14 J of [heat] energy will be released for each '*ejected*' neutron.

In the event a neutron is *actually* ejected, as a proton (alpha) and an electron (beta), they will both achieve an ejection velocity of:

$v = \sqrt{[\ 2.E\ /\ (m_p + m_e)\ ]}$

where $(m_p + m_e)$ = the mass of the electron and the proton

When the ejected proton impacts a neutron in another atom, it cannot impact another proton because of the similar [positive]) charge polarities, the character of impacted atom will alter (Z+1), which can cause natural molecules to fail.

4 Calculation Results

A collection of [mostly] tabulated calculation results for selected examples using the formulas provided in section 3.

4.1 Energy

The following Table lists the energies prevalent in a proton-electron pair at the time of the creation of a neutron.

	Proton	Electron	units
e	2.75902141E-16#	1.6021765E-19	C
m	1.67262164E-27	9.1093897E-31	(kg)
KE	0	4.09355561E-14	J
PE	-8.18711122E-14		J
E	-4.09355561E-14		J

Table 4.1.3-1: Neutron Energies
all of this electrical charge except 'e' is used to generate electro-magnetic energy

The following Table shows the properties of electro-magnetic radiation emitted by a proton-electron pair immediately prior to neutron creation

	Property Value	units
T	623316124.71718	K
f	1.69320448260835E+22	Hz
λ	1.77056263481051E-14	m
A	2.817937953839E-15 #	m
Spectra	γ	

Table 4.1.3-1: Electro-Magnetic Energy
T = temperature; f = frequency; λ = wavelength; A = amplitude
# the neutronic radius (R_n) is 2.81793795383896E-15 m

4.2   Proton-Electron Pair

All electrons in an atom must obey Newton's laws of orbital motion, and their spacing between adjacent electrons in the same and adjacent shells is maintained by Coulomb's force-law. These conditions define the amount of electro-magnetic energy any given electron is able to absorb.

Each shell can hold up to two identical electrons, both of which will absorb the same amount of electro-magnetic energy (heat or temperature) from their surroundings.

This means that the electro-magnetic energy (heat or temperature) absorbed in each shell will be different from each of the other shells. Moreover, in accordance with Newton's laws of motion, electron(s) in the innermost shell will absorb the most energy and the outermost shell will absorb the least.

The heat we feel from matter is the sum of the electro-magnetic energy radiated by each proton-electron pairing within an atom. Its temperature is that generated by the proton-electron pairs in the innermost shell.

The Tables in this Chapter list electron orbital performance at the maximum possible temperature; when the orbiting electron achieves *light-speed.*

Below is listed the descriptions of the additional symbols used in the *electron* Tables.

Tables -1 to -4: (specific to the atom)

T = the temperature of the electron
X = velocity heat coefficient [5]
X_R = radial heat coefficient [5]

Tables -1 to -4 below show the orbital properties of a proton-electron pair at the maximum possible temperature (when the neutron is created).

Sym	Newton	Coulomb	units
T	623316124.7171790	623316124.7171790	K
m_2	9.1093897E-31	9.1093897E-31	kg
X		6.9353271647894E-09	$K.s^2/m^2$
X_R		1.75646616508036E-06	$K.m$

Table 4.2-1: *Input Data* (T_n)

Sym	Newton	Coulomb	units
R	2.817937953839E-15	2.817937953839E-15	m
d	8.852813174052E-15	8.852813174052E-15	m
ℓ	2.817937953839E-15	2.817937953839E-15	m
a	2.817937953839E-15	2.817937953839E-15	m
e	0	0	
b	2.817937953839E-15	2.817937953839E-15	m
p	2.817937953839E-15	2.817937953839E-15	m
f	2.817937953839E-15	2.817937953839E-15	m
x'	0	0	m
A	2.494667824141E-29	2.494667824141E-29	m²
L	1.770562634810E-14	1.770562634810E-14	m
K	0.15587874533403	0.15587874533403	s²/m³

Table 4.2-2: Orbital Shape (T_n)

Sym	Newton	Coulomb	units
m_1	1.67262163783E-27	1.6726216378300E-27	kg
m_2	9.1093897E-31	9.1093897E-31	kg

Table 4.2-3: Masses (T_n)

Sym	Newton	Coulomb	units
v	6.2938005855237E-12	299792459	m/s
g	1.4057061035135E-08	3.189407288078E+31	m/s²
F	1.2805124700573E-38	29.05355389912620	N
Fc	1.2805124700573E-38	29.05355389912620	N
PE	-3.6084046897386E-53	-8.1871112226254E-14	J
KE	1.8042023448693E-53	4.0935556113127E-14	J
E	-1.8042023448693E-53	-4.0935556113127E-14	J
h	1.7735539543841E-26	8.4479654849081E-07	m²/s
PE/KE	-2	-2	

Table 4.2-4: Orbital Performance (T_n)

4.3   The Atom

The relationship between the two primary ratios; mass and velocity (ξ_m & ξ_v respectively – refer to Chapter 5.3) are instrumental in the reason why a proton and its orbiting electron unite at orbital radius 'R$_n$' [3]

The following lists the symbol descriptions for Tables -1 to -4:

Tables -1: *Input Data*

T = *electron temperature*
m_2 = *electron mass*
X = *Heat coefficient for electron velocity and temperature*
X_R = *Heat coefficient for electron orbital radius and temperature*

Tables -2: Orbital Shape

R = radial distance between the centres of the force-centre and the satellite
d = arc distance between equi-spaced points on the surface of a sphere
ℓ = straight-line distance between equi-spaced points on the surface of a sphere
a & b = major and minor orbital semi-axes
e = orbital eccentricity
p = orbital half parameter
f = distance between orbital 'focus' and satellite (R^p)
x' = distance between orbital 'focus' and orbit centre (a - f)
A = total swept area of orbit
L = circumferential length of orbit
K = constant of proportionality

Tables -3: Masses

m_1 = force-centre mass
m_2 = *satellite mass*

Tables -4: Orbital Performance

v = satellite curvilinear velocity
g = gravitational acceleration between force-centre and satellite
F = gravitational force between force-centre and satellite
F_c = centrifugal force on electron
PE = gravitational energy between force-centre and satellite
KE = kinetic energy in satellite
E = total energy should always be the same, irrespective of radial distance
h = constant of motion should always be the same, irrespective of radial distance
PE/KE = confirmation of 2:1 relationship in circular orbits

4.3.1 Cold ($\underline{T}_c$)

Tables -1 to -4 below show the orbital properties of a proton-electron pair at the minimum temperature.

Property	Newton	Coulomb	units
T	0.122060237421696	0.122060237421696	K
m_2	9.1093897E-31	9.1093897E-31	kg
X		6.9353271647894E-09	$K.s^2/m^2$
X_R		1.75646616508036E-06	$K.m$

Table 4.3.1-1: *Input Data* ($\underline{T}_c$)

Property	Newton	Coulomb	units
R	1.439015851667E-05	1.439015851667E-05	m
d	1.439015851667E-05	1.439015851667E-05	m
ℓ	4.520801627996E-05	4.520801627996E-05	m
a	1.439015851667E-05	1.439015851667E-05	m
e	0	0	
b	1.439015851667E-05	1.439015851667E-05	m
p	1.439015851667E-05	1.439015851667E-05	m
f	1.439015851667E-05	1.439015851667E-05	m
x'	0	0	m
A	6.505505204927E-10	6.505505204927E-10	m^2
L	9.041603255991450E-05	9.041603255991450E-05	m
K	0.1558787453340300	0.1558787453340300	s^2/m^3

Table 4.3.1-2: Orbital Shape ($\underline{T}_c$)

Property	Newton	Coulomb	units
m_1	1.67262163783E-27	1.6726216378300E-27	kg
m_2	9.1093897E-31	9.1093897E-31	kg

Table 4.3.1-3: Masses ($\underline{T}_c$)

Property	Newton	Coulomb	units
v	8.8073631286363E-17	4195.2092599071500	m/s
g	5.3904649618566E-28	1.223042867389E+12	m/s^2
F	4.9103846001748E-58	1.1141174098854E-18	N
Fc	4.9103846001748E-58	1.1141174098854E-18	N
PE	-7.0661212774321E-63	-1.6032326134430E-23	J
KE	3.5330606387160E-63	8.0161630672150E-24	J
E	-3.5330606387160E-63	-8.0161630672150E-24	J
h	1.2673935153494E-21	6.0369726260658E-02	m^2/s
PE/KE	-2	-2	

Table 4.3.1-4: Orbital Performance ($\underline{T}_c$)

4.3.2 Minimum Neutron (T_0)

Tables -1 to -4 below show the orbital properties of a proton-electron pair at Planck's minimum temperature [1].

Property	Newton	Coulomb	units
T	210.19332853584	210.19332853584	K
m_2	9.1093897E-31	9.1093897E-31	kg
X		6.9353271647894E-09	$K.s^2/m^2$
X_R		1.75646616508036E-06	$K.m$

Table 4.3.2-1: *Input Data* (T_0)

Property	Newton	Coulomb	units
R	8.356431563816E-09	8.356431563816E-09	m
d	2.625250401111E-08	2.625250401111E-08	m
ℓ	8.356431563816E-09	8.356431563816E-09	m
a	8.356431563816E-09	8.356431563816E-09	m
e	0	0	
b	8.356431563816E-09	8.356431563816E-09	m
p	8.356431563816E-09	8.356431563816E-09	m
f	8.356431563816E-09	8.356431563816E-09	m
x'	0	0	m
A	7.396196699443E-10	7.396196699443E-10	m^2
L	9.640713088869E-05	9.640713088869E-05	m
K	0.15587874533403	0.15587874533403	s^2/m^3

Table 4.3.2-2: Orbital Shape (T_0)

Property	Newton	Coulomb	units
m_1	1.67262163783E-27	1.6726216378300E-27	kg
m_2	9.1093897E-31	9.1093897E-31	kg

Table 4.3.2-3: Masses (T_0)

Property	Newton	Coulomb	units
v	3.6548390907795E-15	174090.86662108400	m/s
g	1.5985111201450E-21	3.6268626876711E+18	m/s^2
F	1.4561460733184E-51	3.3038505610385E-12	N
Fc	1.4561460733184E-51	3.3038505610385E-12	N
PE	-1.2168185008604E-59	-2.7608401110393E-20	J
KE	6.0840925043021E-60	1.3804200555196E-20	J
E	-6.0840925043021E-60	-1.3804200555196E-20	J
h	3.0541412738857E-23	1.4547784128045E-03	m^2/s
PE/KE	-2	-2	

Table 4.3.2-4: Orbital Performance (T_0)

4.3.3   Planck Mean ($\underline{T}_m$)

Tables -1 to -4 below show the orbital properties of a proton-electron pair at Planck's mean temperature.

Property	Newton	Coulomb	units
T	361962.55467156	361962.55467156	K
m_2	9.1093897E-31	9.1093897E-31	kg
X		6.9353271647894E-09	$K.s^2/m^2$
X_R		1.75646616508036E-06	$K.m$

Table 4.3.3-1: Input Data ($\underline{T}_m$)

Property	Newton	Coulomb	units
R	4.852618433623E-12	4.852618433623E-12	m
d	1.524495042174E-11	1.524495042174E-11	m
ℓ	4.852618433623E-12	4.852618433623E-12	m
a	4.852618433623E-12	4.852618433623E-12	m
e	0	0	
b	4.852618433623E-12	4.852618433623E-12	m
p	4.852618433623E-12	4.852618433623E-12	m
f	4.852618433623E-12	4.852618433623E-12	m
x'	0	0	m
A	2.193772531476E-16	2.193772531476E-16	m^2
L	5.250500802222E-08	5.250500802222E-08	m
K	0.15587874533403	0.15587874533403	s^2/m^3

Table 4.3.3-2: Orbital Shape ($\underline{T}_m$)

Property	Newton	Coulomb	units
m_1	1.67262163783E-27	1.6726216378300E-27	kg
m_2	9.1093897E-31	9.1093897E-31	kg

Table 4.3.3-3: Masses ($\underline{T}_m$)

Property	Newton	Coulomb	units
v	1.5166683358448E-13	7224342.80705005	m/s
g	.7402920143405E-15	1.0755250944965E+25	m/s^2
F	4.3181167250426E-45	9.7973772178982E-06	N
Fc	4.3181167250426E-45	9.7973772178982E-06	N
PE	-2.0954172818476E-56	-4.7542933288727E-17	J
KE	1.0477086409238E-56	2.3771466644364E-17	J
E	-1.0477086409238E-56	-2.3771466644364E-17	J
h	7.3598127242123E-25	3.5056979076300E-05	m^2/s
PE/KE	-2	-2	

Table 4.3.3-4: Orbital Performance ($\underline{T}_m$)

4.3.4 Neutron ($\underline{T}_n$)

Tables -1 to -4 below show the orbital properties of a proton-electron pair at the maximum possible temperature (when the neutron is created).

Property	Newton	Coulomb	units
T	623316124.71718	623316124.71718	K
m_2	9.1093897E-31	9.1093897E-31	kg
X		6.9353271647894E-09	$K.s^2/m^2$
X_R		1.75646616508036E-06	$K.m$

Table 4.3.4-1: *Input Data* ($\underline{T}_n$)

Property	Newton	Coulomb	units
R	2.817937953839E-15	2.817937953839E-15	m
d	8.852813174052E-15	8.852813174052E-15	m
ℓ	2.817937953839E-15	2.817937953839E-15	m
a	2.817937953839E-15	2.817937953839E-15	m
e	0	0	
b	2.817937953839E-15	2.817937953839E-15	m
p	2.817937953839E-15	2.817937953839E-15	m
f	2.817937953839E-15	2.817937953839E-15	m
x'	0	0	m
A	2.494667824141E-29	2.494667824141E-29	m²
L	1.770562634810E-14	1.770562634810E-14	m
K	0.15587874533403	0.15587874533403	s²/m³

Table 4.3.4-2: Orbital Shape ($\underline{T}_n$)

Property	Newton	Coulomb	units
m_1	1.67262163783E-27	1.6726216378300E-27	kg
m_2	9.1093897E-31	9.1093897E-31	kg

Table 4.3.4-3: Masses ($\underline{T}_n$)

Property	Newton	Coulomb	units
v	6.2938005855237E-12	299792459	m/s
g	1.4057061035135E-08	3.189407288078E+31	m/s²
F	1.2805124700573E-38	29.05355389912620	N
Fc	1.2805124700573E-38	29.05355389912620	N
PE	-3.6084046897386E-53	-8.1871112226254E-14	J
KE	1.8042023448693E-53	4.0935556113127E-14	J
E	-1.8042023448693E-53	-4.0935556113127E-14	J
h	1.7735539543841E-26	8.4479654849081E-07	m²/s
PE/KE	-2	-2	

Table 4.3.4-4: Orbital Performance ($\underline{T}_n$)

The following Table is a check-list of the ratios between various Newton (^{N}) and Coulomb (^{C}) calculated properties:

Property Ratio		Value
$v^N : v^C$	$\sqrt{\varphi}$	2.0993858906650E-20
$g^N : g^C$	φ	4.40742111792335E-40
$F^N : F^C$	φ	4.40742111792335E-40
$Fc^N : Fc^C$	φ	4.40742111792335E-40
$PE^N : PE^C$	φ	4.40742111792335E-40
$KE^N : KE^C$	φ	4.40742111792335E-40
$E^N : E^C$	φ	4.40742111792335E-40
$h^N : h^C$	$\sqrt{\varphi}$	2.09938589066502E-20

All of which are either the coupling ratio or its square-root

4.4   Isotope

The following list provides the nominal neutronic ratio (ψ) of the first 94 elements in the Periodic Table:

Element	Z	ψ	Element	Z	ψ	Element	Z	ψ
Hydrogen	1	0.0	Germanium	32	1.27	Gadolinium	64	1.46
Helium	2	1.0	Arsenic	33	1.27	Terbium	65	1.45
Lithium	3	1.31	Selenium	34	1.32	Dysprosium	66	1.46
Beryllium	4	1.25	Bromine	35	1.28	Holmium	67	1.46
Boron	5	1.16	Krypton	36	1.33	Erbium	68	1.46
Carbon	6	1.0	Rubidium	37	1.31	Thulium	69	1.45
Nitrogen	7	1.0	Strontium	38	1.31	Ytterbium	70	1.47
Oxygen	8	1.0	Yttrium	39	1.28	Lutetium	71	1.46
Fluorine	9	1.11	Zirconium	40	1.28	Hafnium	72	1.48
Neon	10	1.02	Niobium	41	1.27	Tantalum	73	1.48
Sodium	11	1.09	Molybdenum	42	1.28	Tungsten	74	1.48
Magnesium	12	1.03	**Technetium**	**43**	**1.3**	Rhenium	75	1.48
Aluminium	13	1.08	Ruthenium	44	1.3	Osmium	76	1.5
Silicon	14	1.01	Rhodium	45	1.29	Iridium	77	1.5
Phosphorus	15	1.06	Palladium	46	1.31	Platinum	78	1.5
Sulphur	16	1.0	Silver	47	1.3	Gold	79	1.49
Chlorine	17	1.09	Cadmium	48	1.34	Mercury	80	1.51
Argon	18	1.22	Indium	49	1.34	Thallium	81	1.52
Potassium	19	1.06	Tin	50	1.37	Lead	82	1.53
Calcium	20	1.0	Antimony	51	1.39	Bismuth	83	1.52
Scandium	21	1.14	Tellurium	52	1.45	Polonium	84	1.49
Titanium	22	1.18	Iodine	53	1.39	**Astatine**	**85**	**1.47**
Vanadium	23	1.21	Xenon	54	1.44	**Radon**	**86**	**1.56**
Chromium	24	1.17	Caesium	55	1.42	**Francium**	**87**	**1.56**
Manganese	25	1.2	Barium	56	1.45	**Radium**	**88**	**1.57**
Iron	26	1.15	Lanthanum	57	1.44	**Actinium**	**89**	**1.55**
Cobalt	27	1.18	Cerium	58	1.42	**Thorium**	**90**	**1.58**
Nickel	28	1.1	Praseodymium	59	1.39	**Protactinium**	**91**	**1.57**
Copper	29	1.19	Neodymium	60	1.4	**Uranium**	**92**	**1.59**
Zinc	30	1.18	Promethium	61	1.38	**Neptunium**	**93**	**1.55**
Gallium	31	1.25	Samarium	62	1.43	**Plutonium**	**94**	**1.6**
			Europium	63	1.41			

The elements highlighted in **bold type** are those that are naturally radioactive.

Plutonium and the other elements with a greater atomic number will readily divide into smaller atoms.

5 The Physical Constants

All the physical constants (including electrical properties such as Volts, Amps, Henries, Farads, Ohms, etc.) are provided (to ≤ 15 decimal places) in terms of the same four basic units; length, time, mass and charge and two ratios: m_e, e, R_n, t_n & ξ_v, ξ_m

5.1 Introduction

The physical constants, are calculated here from just four fundamental constants (mass, charge, length, time) and two ratios (mass and velocity): m_e, e, R_n, t_n & ξ_m, ξ_v

Because we have (to date) concentrated on defining nature in terms of electricity, magnetism or mechanics, the relationship between these properties has become very difficult to reconcile mathematically. For example; we cannot readily explain Volts, Amps, Ohms, Henries, Farads, etc., in terms of energy.

In this book, I have therefore reduced everything to the same basic [metric] units (Imperial conversions are in parenthesis):

Magnetic Charge: kilogram (kg)
(1kg = 2.20462262lb)

Electrical Charge: Coulomb (C)
(1C = 1C)

Distance: metre (m)
(1m = 39.3700787401575in)

Time: second (s)
(1s = 1s)

Notes:
1) Joules and Newtons remain useful, but they are merely compilations of the above.
2) Converting to imperial units ...
... between numerators or denominators: multiply by the conversion factor above
... across numerators and denominators: divide by the conversion factor above

These are the fundamental units that define all others. *Everything* can be explained and described (mathematically) using them.

All formulas for the constants provided in this section are described and explained in "PHILOSOPHIÆ NATURALIS PRINCIPIA MATHEMATICA Revision IV". The relationship between mass and magnetic charge is also explained in the same publication.

5.2   Symbols

The following is an alphabetical list of the symbols explained in the Tables (in this Chapter 5) indicated; **Table 5.?**.

Those that are new, i.e. not currently known, are highlighted in **bold text**

Symbol	Description	Table(s)
a_o	Rydberg radius (also known as Bohr Radius)	4
A	electrical current	8
c	speed of electro-magnetic radiation	4
$C_?$	specific heat capacity	6
$C_?$	heat capacity	5, 6
e	elementary charge unit	3
e	natural logarithm	4
E	energy	
F	Farad	8
F	Force	
G	Newton's gravitational constant	4
h	Planck's constant	4
ℏ	Planck's constant (Dirac version)	4
h'	modified Planck's constant	4
H	Henry	8
k	Coulomb's constant	4
k'	Coulomb's constant (modified)	4
k_B	Boltzmann's constant	5
K	Constant of proportionality	Chapter 3.2.2
m	mass	3, 4
m_e	mass of an electron	3
m_p	mass of a proton	4
m_n	mass of a neutron	4
$N_?$	microstate	5
N_A	Avogadro's number	
q	specific charge capacity	7
Q	charge capacity	5, 7

Table 5.2a

Symbol	Description	Table(s)
r	particle (or body) radius	
R	orbital radius	4.3.1 to 4
$R_?$	gas constant	5
R_a	specific gas constant	6
RAC	relative atomic charge	7
RAM	relative atomic mass	6
$\mathbf{R_c}$	charge [emission] capacity	5, 7
RC	relative charge capacity	5
R_i	ideal gas constant	5
$\mathbf{R_n}$	neutronic radius	4
$\mathbf{R_p}$	relative charge capacity (constant pressure)	5
$\mathbf{R_T}$	gas constant (temperature dependent)	5
R_∞	Rydberg's wave number	4
R_γ	Rydberg's universal constant (energy)	4
t	time	3
$\mathbf{t_n}$	neutronic period	3
$T_?$	Temperature (key)	4.3.1 to 4
$\mathbf{T_n}$	neutronic temperature	3
v	velocity	4.3.1 to 4
V	electrical voltage	8
V	volume	
X	heat coefficient (velocity)	5
$\mathbf{X_R}$	heat coefficient (orbital radius)	5
Y	temperature coefficient	4
ε_o	permittivity of a vacuum	4
λ	wavelength	3.1.3
μ', μ_o	magnetic constant	4
ρ	density	
Σ	universal constant	3
$\boldsymbol{\varphi}$	coupling ratio	4
Ω	electrical resistance	8
ξ_m	mass ratio	4
ξ_v	velocity ratio	4

Table 5.2b

Symbol	Description
Suffix:	
e	electron
n	neutronic
n	atomic shell number
p	proton
p	constant pressure (heat & charge capacity)
t	constant temperature (heat & charge capacity)
u	ultimate
v	constant volume (heat & charge capacity)
Atomic: temperature, velocity & orbital radius:	
c	cold
m	mean Planck value
n	neutronic
o	minimum Planck value
Modifier:	
N	Newton
P	Planck

Table 5.2c

5.3   Primary Constants

The Tables in this Chapter provide the meaning of the symbols previously listed in Table 2.1 In the following Chapters, the constants are grouped according to their properties; primary, general, heat, charge, etc.

All new constants are highlighted in **bold text**

There are very few primary constants, i.e. those that we must take for granted and from which all other physical variables and constants can be calculated. These are listed below

Symbol	Value	Units
e	1.60217648753E-19	C
electrical charge (elementary charge unit)		
m_e	9.1093897E-31	kg
magnetic charge (the *mass* of an electron)		
R_n	2.81793795383896E-15	m
distance (the neutronic radius)		
t_n	5.90596121302193E-23	s
time (neutronic period)		
$\S_m$	1836.15115053207	
static ratio $\{m_p/m_e\}$		
$\S_v$	1722.0458764934	
dynamic ratio $\{c/v_o\}$		
Σ	3E-91 (exact)	m^6
universal constant		
T_n	623316124.717178	K
temperature (neutronic temperature)		
Table 2.2		

Notes
Temperature is not a real property. It is an interpretive value for the kinetic energy of the electron(s) orbiting in an atom's innermost shell.

5.4 General Physical Constants

Symbol	Formula	Value	Units
G	$a_o.c^2 / \boldsymbol{\rho_u}$	6.67359232004334E-11	$m^3 / s^2.kg$
Newton's gravitational constant (per m^3)			
k	$c^2.\mu'$	8.98755184732667E+09	$J.m / C^2$
Coulomb's constant [for an electron] (refer to Chapter 6.11.4)			
k'	$k / \xi_m{}^2$	2.6657815048876E+03	$J.m / C^2$
Coulomb's constant for a proton			
φ	$G.m_e.m_p / k.e^2$	4.40742111792334E-40	
Coupling Ratio			
μ'	$\mathbf{R_n}.m_e/e^2$	1E-07	$kg.m / C^2$
Magnetic constant (fundamental)			
μ_o	$4\pi.\mu'$	1.25663706143592E-06	$kg.m / C^2$
Magnetic constant			
ε_o	$1 / \mu_o.c^2$	8.85418775855161E-12	$C^2 / J.m$
Permittivity of a vacuum (e.g. within an atom)			
h	$\frac{1}{2}.\mathbf{R_n}.m_e.c.\xi_v$	6.62607174469163E-34	$kg.m^2/s$
Planck's constant (resolved into its component parts)			
ħ	$h / 2\pi$	1.05457207144921E-34	$kg.m^2/s$
Planck's constant (modified by Dirac)			
h'	$\frac{1}{2}.\mathbf{R_n}.m_e.c^2$	1.15353857232684E-28	$J.m$
Modified Planck's constant			
R_∞	$1 / a_o.\xi_v$	1.09737269561359E+07	$/m$
Rydberg's wave number			
R_y	$\mathbf{R_n}/a_o . \frac{1}{2}.m_e.c^2$	2.17987197684936E-18	J
Rydberg's universal constant for the energy of an electron			
α	$e^2 / 4\pi$	2.04272942122269E-39	C^2
Fine structure constant			
X	T_n/c^2	6.9353271647894E-09	$K.s^2/m^2$
Velocity constant			
X$_R$	$T_n.R_n$	1.75646616508035E-06	$K.m$
Radial constant			
Y	$\sqrt[3]{[\frac{1}{2}.\xi_v]}$	9.51345439232503	
Temperature coefficient			
e'	$e.\xi_v.\sqrt{[T/T_n]}$	2.75902141376572E-16	
Proton charge			

Table 5.4a

Atomic property constants (refer to Chapter 6.10 for particle properties):

Symbol	Formula	Value	Units
m_e			kg
Mass of an electron (refer to Chapter 5.3)			
m_p	$m_e.\xi_m$	1.672621637830E-27	kg
Mass of a proton			
m_n	m_e+m_p	1.6735325768E-27	kg
Mass of a neutron			
a_o	$R_n.(\xi_v/4\pi)^2$	5.2917721067E-11	m
Rydberg's radius			
R_c	$R_n.\xi_v^3$	1.43901585166681E-05	m
Cold orbital radius (refer to Chapter 3.3.4)			
R_o	$R_n.\xi_v^2$	8.3564315638157E-09	m
Planck minimum orbital radius (refer to Chapter 3.3.4)			
R_m	$R_n.\xi_v$	4.85261843362263E-12	m
Planck mean orbital radius (refer to Chapter 3.3.4)			
R_n			m
Neutronic radius (refer to Chapter 3.3.4)			
v_c	$v_o . \sqrt{[R_o/R_c]}$	4195.20925990715	m/s
Electron cold velocity (refer to Chapter 3.3.4)			
v_o	$c . \sqrt{[R_n/R_o]}$	174090.866621084	m/s
Electron minimum Planck orbital velocity (refer to Chapter 3.3.4)			
v_m	$\sqrt{[c.v_o]}$	7224342.80705004	m/s
Electron mean Planck orbital velocity (refer to Chapter 3.3.4)			
c	$2\pi.R_n / t_n$	299792459	m/s
Electron neutronic velocity (refer to Chapter 3.3.4)			
T_c	$X.v_c^2$	0.122060237421696	K
Cold temperature (refer to Chapter 3.3.4)			
T_o	$X.v_o^2$	210.193328535837	K
Planck minimum temperature (refer to Chapter 3.3.4)			
T_m	$X.v_m^2$	361962.554671561	K
Planck mean temperature (refer to Chapter 3.3.4)			
T_n	$X.c^2$		K
Neutronic temperature (refer to Chapter 3.3.4)			
e	$\exp(1)$	2.71828182845905	
Natural logarithm			

Table 5.4b

5.5 Universal Heat & Charge Capacities

Symbol	Formula	Value	Units
k_B	$m_e.c^2 / Y.\underline{T}_n$	1.38065156E-23	J/K
Boltzmann's constant			
R_i	$k_B.N_A$	8.31447876657891	J / K.mol
Ideal gas constant			
RC	e/m_e	1.75881869180545E+11	C/kg
Relative charge capacity			
R_c	$\sqrt{[\,G/k\,]}$	8.61706029887134E-11	C/kg
Charge [emission] capacity			
R_a	R_i / RAM		J / kg.K
Specific gas constant			
R	$m.R_a$	1.38065156E-23	J/K
Gas constant			
R_p	$c_p.RAM$	20.7861969164473	J / K.mol
Gas constant; R_i multiplied by 2.5			
R_T	$RAC.q_p.Ln(\underline{T})$ $R_i.Ln(N_t)$		J / K.mol
Gas constant ($R_T = R_i$ when $N_t = e$ & $\underline{T}$ = 1.49182469764127 K)			
C_t	$m.c_t$		J/K
Heat capacity (constant *temperature*)			
C_V	$m.c_V$		J/K
Heat capacity (constant *volume*); C_t multiplied by 1.5			
C_p	$m.c_p$		J/K
Heat capacity (constant *pressure*); C_t multiplied by 2.5			
Q_t	$e.q_t$		J/K
Charge capacity (constant *temperature*); also equal to **R** & C_t			
Q_V	$e.q_V$		J/K
Charge capacity (constant *volume*); Q_t multiplied by 1.5; also equal to C_V			
Q_p	$e.q_p$		J/K
Charge capacity (constant *pressure*); Q_t multiplied by 2.5; also equal to C_p			

Table 5.5a

5.5.1 Microstates

Symbol	Formula	Value	Units
N_t	$\exp(c_p.L_n(T) / R_a)$ $\exp(\mathbf{q_p}.L_n(T) / R_a)$ $\exp(2.5 \cdot Ln(T))$		
Microstate (constant *temperature*)			
N_V	c_v / R_a $\mathbf{q_v} / R_a$		
Microstate (constant *volume*); N_t multiplied by 1.5			
N_p	c_p / R_a $\mathbf{q_p} / R_a$		
Microstate (constant *pressure*); N_t multiplied by 2.5			
Table 5.5b			

5.6   Specific Heat Capacities (particles)

Symbol	Formula	Value	Units
RAM$_e$	$m_e.N_A$	5.4858031839070700E-07	kg/mol
Relative atomic mass of an electron			
RAM$_p$	$m_p.N_A$	1.00727638277235E-03	kg/mol
Relative atomic mass of a proton (also the RAM of an hydrogen atom)			
R_{ae}	R_i / RAM$_e$ = k_B / m_e	1.51563563034308E+07	J / kg.K
Specific gas constant for an electron			
R_{ap}	R_i / RAM$_p$ = k_B / m_p	8.25441647276088E+03	J / kg.K
Specific gas constant for a proton			
c_{et}	k_B / m_e	1.51563563034305E+07	J / kg.K
Specific heat capacity for the electron (constant *temperature*)			
c_{eV}	$1.5 . c_{et}$	2.27345344551458E+07	J / kg.K
Specific heat capacity for the electron (constant *volume*)			
c_{ep}	$c_{et} + c_{eV}$	3.78908907585763E+07	J / kg.K
Specific heat capacity for the electron (constant *pressure*)			
c_{pt}	k_B / m_p	8.25441647276074E+03	J / kg.K
Specific heat capacity for the proton (constant *temperature*)			
c_{pV}	$1.5 . c_{pt}$	1.23816247091411E+04	J / kg.K
Specific heat capacity for the proton (constant *volume*)			
c_{pp}	$c_{pt} + c_{pV}$	2.06360411819018E+04	J / kg.K
Specific heat capacity for the proton (constant *pressure*)			
C_t	$m_e.c_{et}$ $m_p.c_{pt}$	1.38065156E-23	J/K
Heat capacity (constant *temperature*); equal to **R** & **Q$_t$**			
C_V	$m_e.c_{eV}$ $m_p.c_{pV}$	2.07097734E-23	J/K
Heat capacity (constant *volume*); equal to **Q$_v$**			
C_p	$m_e.c_{ep}$ $m_p.c_{pp}$	3.4516289E-23	J/K
Heat capacity (constant *pressure*); equal to **Q$_p$**			

Table 5.6

5.7 Specific Charge Capacities (particles)

Symbol	Formula	Value	Units
RAC_e	$e.N_A$	96485.3317942158	C/mol
Relative atomic charge of an electron (also equal to the Farad)			
RAC_p	$e'.N_A$	1.77161652983418E+08	C/mol
Relative atomic charge of a proton (also the RAC of an hydrogen atom)			
R_{ce}	R_i / RAC_e	8.61735002820125E-05	J / C.K
Specific gas constant for an electron			
R_{cp}	R_i / RAC_p	4.69315939796359E-08	J / C.K
Specific gas constant for a proton			
q_{et}	k_B / m_e	8.61735002820123E-05	J / C.K
Specific charge capacity for the electron (constant *temperature*)			
q_{ev}	$1.5 . q_{et}$	1.29260250423019E-04	J / C.K
Specific charge capacity for the electron (constant *volume*)			
q_{ep}	$q_{et} + q_{ev}$	2.1543375070503E-04	J / C.K
Specific charge capacity for the electron (constant *pressure*)			
q_{pt}	k_B / e'	4.69315939796358E-08	J / C.K
Specific charge capacity for the proton (constant *temperature*)			
q_{pv}	$1.5 . q_{pt}$	7.0397390969454E-08	J / C.K
Specific charge capacity for the proton (constant *volume*)			
q_{pp}	$q_{pt} + q_{pv}$	1.1732898494909E-07	J / C.K
Specific charge capacity for the proton (constant *pressure*)			
Q_t	$e.q_{et}$ $e'.q_{pt}$	1.38065156E-23	J/K
Charge capacity (constant *temperature*); equal to **R** & C_t			
Q_v	$e.q_{ev}$ $e'.q_{pv}$	2.07097734E-23	J/K
Charge capacity (constant *volume*); equal to C_v			
Q_p	$e.q_{ep}$ $e'.q_{pp}$	3.4516289E-23	J/K
Charge capacity (constant *pressure*); equal to C_p			

Table 5.7

5.8 Electricity

Apart from the Farad, no values are provided for the following electrical properties because the Amp, Volt, Ohm and Henry are now redundant.

Symbol	Formula	Value	Units
A	e.f		C/s
Electrical current (Coulomb flow-rate)			
V	PE/e		J/C
Electrical voltage (potential energy per coulomb)			
Ω	V/A = PE / f.e^2		J.s/C^2
Electrical resistance (momentum over distance per Coulomb squared)			
H	μ_o.R		kg.m^2/C^2
Henry; unit of mutual inductance			
F	e.N$_A$	96485.3317942156	C/mol
Farad; unit of electrostatic capacitance (equal to **RAC$_e$**)			
P	V.A = PE.f		J/s
Power (Watt)			

Table 5.8

PE is the potential energy between a proton and its orbiting electron

6 The Laws of Thermodynamics

The First Law of Thermodynamics: Conservation of energy
Energy can never be lost, it can only be transformed or transferred.

The Second Law of Thermodynamics: Heat will not spontaneously pass from a colder body to hotter body

A high-energy source (hotter body) will spontaneously lose energy to a low-energy source (colder body) but you must add work if you want energy to transfer in the other direction (up-hill so to speak). This law essentially states that it is impossible to create energy from nothing.

This law also claims that energy can, and in fact is, lost by a system to its surroundings but that the reverse cannot happen i.e. an increase in disorder is an inevitable feature of time.

The Third Law of Thermodynamics: The entropy of a substance approaches zero as its temperature approaches zero (absolute)

Entropy is the term used to define disorder. The higher a substance's temperature the more disordered will be its atomic structure and the higher its entropy. E.g. gas has a higher entropy than a solid substance.

7 Model Verification

Apart from the correlation between PVRT and the gas-pressure calculation method using the model (refer to Chapter 3.3.3), the following chapters provide additional supporting mathematical verification of the atomic model described in this publication.

7.1   Density vs Temperature

The magnetic field energy (MFE) generated by proton-electron pairs holds adjacent atoms together (viscosity) and is constant irrespective of temperature. The electrical charges (EC) generated in protons repel adjacent atoms (gases) and varies between e and e′ with temperature.

Given that temperature effects (e.g. gasification) are dependent upon repulsive EC between adjacent atoms and density is dependent upon an attractive MFE between those same atoms, and that both charges are created by the same energy generation process (proton-electron pairs), density and temperature should follow similar patterns of behaviour according to the number of nucleic protons (atomic number (Z)).

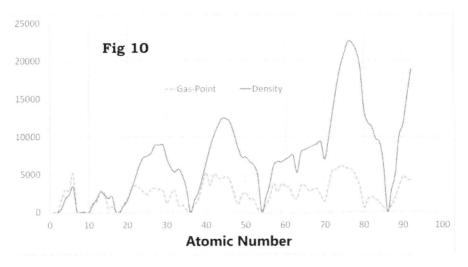

This relationship (between EC and MFE), which can clearly be seen in the temperature/density vs atomic number plot shown in Fig 10 is governed by the nucleic structure, which is the last significant piece of the atomic puzzle. The nucleus, which can be resolved mathematically (refer to Chapter 3.3.2), will be included in a later edition of this publication.

7.2   Specific Heat Capacity

The specific heat capacity of an atom defines the amount of energy it can hold in relation to its mass per unit temperature. This means the sum of the kinetic energy of all electrons in an atom's shells relative to its mass and *'temperature'*.

Temperature in this case (as in all cases) is as defined in the last paragraph of page Chapter 2.2 and an element's *specific heat capacity* is calculated according to the formula provided on page 53 (Chapter 3.3)

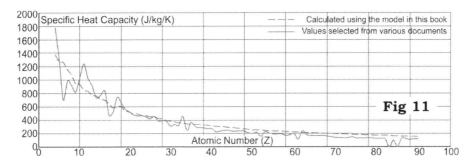

Fig 11 shows the calculated values for specific heat for all atoms from Z=4 to 92 compared to the documented values that have been taken from various sources and which are subject to experimental error.

This calculation technique, is as follows:

$SHC = KE_T \ / \ Y.m.T_1 \quad \{J/kg/K\}$

where:
KE_T = the total kinetic energy in every electron in the elemental shells
T_1 = the temperature of the electron(s) in the innermost shell
Y (refer to Chapter 5.4)
m = the total mass of the atom (including electrons and neutrons)

7.3   Gas-Point

The gas-point of any atom is the temperature at which its electrical charge (EC) exceeds its magnetic field energy (MFE).

If the MFE is greater than the total exposed EC, the atoms will exist as viscous matter; otherwise they will exist as a gas.

Outlying nucleic neutrons protect adjacent atoms from EC. The more outlying neutrons, the greater the protection (higher gas-point temperatures and greater densities).

Density rises with atomic number (Fig 9) because larger atoms tend to collect a greater percentage of neutrons, due to the higher collective MFE within atoms.

Gas-point temperatures also rise with atomic number (Fig 12) for the same reason, but this rise is much less marked because only the outlying and exposed proton EC actively repels.

A mathematical relationship that reflects reality has been identified for all the atoms in the Periodic Table (refer to Chapter 3.3.2).

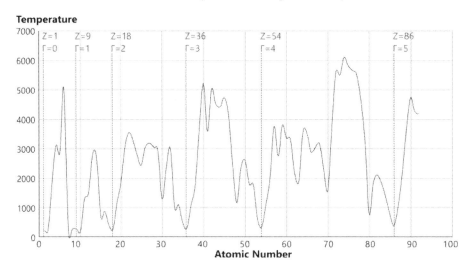

This relationship ($Z \rightarrow \Gamma$) forms the basis for mathematical chemistry.

7.4 Light

Almost all (>99.3%) natural hydrogen exists as lone protons (H+), which cannot be solidified or liquefied due to their similar positive charges. Lone protons also have no way of absorbing or emitting electro-magnetic energy (heat and light).

It is currently claimed that our sun was created from a cloud of hydrogen atoms that accreted into a star due to gravity and an *external force*. This is of course, impossible (refer to Chapter 2.4)

The surface temperature of our sun is said to be about 5778K, which would be impossible if the sun's surface comprised lone protons that cannot collect or emit electro-magnetic energy (i.e. heat or colour).

Yet, the sun's colour and the infra-red light radiated by matter here on Earth (refer to Chapter 2.1.1.2) can be predicted by this atomic model:

Light:	Our Sun	Matter here on Earth
Temperature {K}	5778	303.15
KE {J}	3.74938021543E-19	1.99090210306E-20
electron velocity {m/s}	912757.252	209071.728464087
orbital radius {m}	3.03992067E-10	5.79404969514E-09
f = v / 2πR {Hz}	4.77873747733E+14	5.74292607013E+12
λ = c/f {m}	**6.2734657516211E-07**	**5.220204044751E-05**

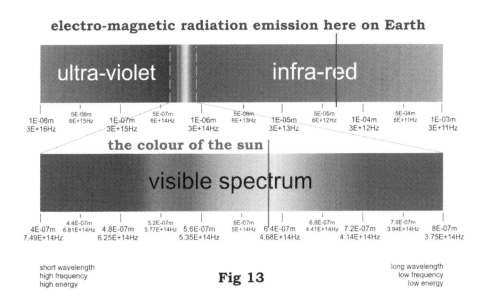

Fig 13

This demonstrates that the hydrogen at the surface of the sun is predominantly proton-electron pairs, which in turn means that these hydrogen atoms (proton-electron pairs) *must be a by-product of fission.*

7.5   PVRT

The final proof of this atomic model can be seen by the replacement of the well-known theory P.V = n.Ri.T with an alternative calculation using the potential energy in a proton-electron pair as described in Chapter 3.3.1

Today, we calculate the pressure (p) of a gas thus:

$$\mathbf{p = n.R_i.T \ / \ V}$$

where: 'n' is the number of moles in the gas, 'R_i' is the ideal gas constant, 'T' is its temperature and 'V' its volume.

But it can also be calculated thus:

$$\mathbf{p = \rho.PE_1 \ / \ m_M.Y} = T.m_e.\rho \ / \ X.Y.m_M = k_B.T.\rho \ / \ m_M$$

where: 'ρ' is the gas density, 'PE_1' is the potential energy between the proton and the electron in shell-1 and 'm_M' is the molecular mass:

$$PE = m_e.v_e^2$$
$$v_e = \sqrt{[T/X]}$$

which provides *exactly* the same result as the PVRT calculation method but is much simpler because there is no need to play with moles.

Because this latter calculation method replaces 'PVRT' altogether, along with the need for Boltzmann's constant, Avogadro's number, gas temperature and the ideal gas constant with potential energy, the model described here must be considered correct.

It is also interesting to note that the lattice structure we know applies to viscous matter (ζ : refer to Chapter 3.3.2) also applies to the same elements in gaseous form, and is responsible for *partial pressure*.

7.6   Logic

Apart from the mathematical evidence (refer to Chapters 7.1 to 7.5), Newton's atom with two electrons per circular orbit is also the most logical atomic configuration:

1) An electro-magnetic particle orbiting another electro-magnetic particle is the only known, practical method of generating electro-magnetic energy

2) A circular orbit is the only shape that will produce steady (non-fluctuating), predictable electro-magnetic energy (Balmer lines)

3) Fig 14 shows that only 2 electrons per orbit can predict the specific heat capacity of all the elements

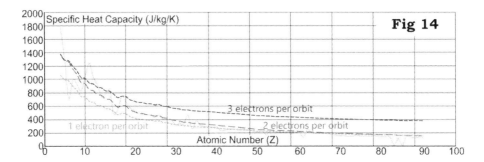

4) 2 electrons per orbit is the obvious orbital arrangement because both are electrically isolated by the positively charged nucleus

5) Exact balance between all electrons in adjacent shells can be mathematically predicted irrespective of orbital position

6) The pressure and temperature in gases can be accurately predicted using proton-electron pair 'PE'; replacing the accepted 'PVRT' method

7) Circular orbits are the only logical explanation for $E=mc^2$

8) '$R_{n'}$' is a genuine (albeit missing until now) natural constant that proves electrons *must* orbit in circular paths

9) Electrical voltage (potential difference) can only be predicted on the basis of circular orbits

10) Newton's atom and Planck's alternative atom can only be reconciled with circular orbits

11) Circular orbits are the only method of reproducing constant, consistent and predictable atomic properties.

Quantum theory must be declared obsolete as none of the above can be applied to its atomic model; items 1) and 8) above being the most damming proof.

8 Things You Can Do

After two and a half years intensive study of this subject I'm knackered! I have left the last bits for somebody else to solve: you perhaps!

[1]   Electron Kinetic Energy

It is not known (by me) whether or not an electron can convert electromagnetic radiation into kinetic energy when in free-flight.

[2]   Nucleic Structure

The mathematical description of nucleic structure is yet to be defined and will be the subject of a future revision of "PHILOSOPHIÆ NATURALIS PRINCIPIA MATHEMATICA Revision IV".

[3]   §m & §v

The relationship between these two ratios is yet to be defined and will be the subject of a future edition of this book.

[4] $\Gamma > \zeta$

Once the mathematical relationship between 'ζ' and 'Γ' has been identified, the atomic puzzle will have been completely solved.

Appendices

References, symbols, glossary, etc. used throughout this book along with a summary list of corollaries and hypotheses.

A1 General

N/A

A2 References

Most of the references used for the creation of this book are from original work supplied in CalQlata (www.calqlata.com), but some additional sources are listed below:

Magnificent Principia; Colin Pask; 978-1-61614-745-7

Seven Brief Lessons on Physics; Carlo Rovelli; 978-0-141-98172-7

Science Data Book; Open University; 0 05 002487 6

Science and Technology Dictionary; Chambers; 0-550-18026-5

A Dictionary of Scientific Units; H G Jerrard & D B McNeill; 0-412-28100-7

It is important to note here that most of the sources here are from work done by pre-20th Century scientists that are universally known and available from sources too numerous to mention here.

A3 Glossary

Alpha-Particle	An ejected proton
Atom	A collection of proton-electron pairs in which all the protons reside inside the same electron shells
Atomic Number	The number of protons in an atom
Atomic Particle	One of the three components that comprise an atom
Beta-Particle	An ejected electron
Big-Bang	The eruption that occurred when the Ultimate-Body accumulated sufficient 'mass' to compromise the integrity of the innermost neutron.
Coupling Ratio (φ)	The ratio of the coupling force due to a magnetic charge and the coupling force due to an electric charge: $\varphi = G.m_p.m_e \div k.e^2$
Deuterium	A proton-electron pair with one neutron partner
Element	An atom that contains at least one proton-electron pair
Gas	Atoms that possess greater electrical field energy than magnetic field energy
Neutronic	The conditions that apply to a proton-electron pair at the instant they become a neutron
Proton-electron pair	A proton that hosts an orbiting electron
Sub-Atomic Particle	The many particles said to compromise atomic particles (leptons, gluons, fermions, quarks, etc.)
Tritium	A proton-electron pair with two neutron partners
Ultimate-Body	A body that contains all the Quanta in the universe ($\approx$2.8E+75) and represents the maximum single 'mass' that can exist without generating a Big-Bang.
Ultimate Density	The mass-density of all three atomic particles $\rho = 7.12660796350449E+16$ kg/m³ Nothing in nature has a 'mass'-density greater than this value
Universal Period	The time elapsed since the last Big-Bang or between subsequent Big-Bangs
Viscous	Solid or liquid matter in which magnetic field energy is greater than electrical field energy

All other definitions can be found on the following web page:
http://calqlata.com/help_definitions.html

A4 Symbols

Refer to Chapter 5 for a list of all the symbols used in this book.

The most prominent subscripts are listed below:

mass	e	electron
	p	proton
temperature	c	cold
	o	minimum Planck
	m	mean Planck
	n	maximum Planck
Rydberg	γ	energy constant
	∞	wave number
	o	orbital radius (a_o) *occasionally referred to as the Bohr radius*
Others	u	Ultimate
radii	n	Neutron orbit radius
	s	Schwarzschild radius
	1	force-centre
	2	satellite
energy	e	electron
	p	proton

The most prominent superscripts are listed below:

Force	N	Newton
	P	Planck

A5 Useful Formulas

Equidistant arc-length between 'n' points on the surface of a sphere:

$d = \pi.A / C.n$

where C is the circumference of the sphere

Linear distance across arc-length 'd' (above):

$\ell = 2.R.Sin(\frac{1}{2}.d/R)$

but if you know 'ℓ' and need to find 'n':

$n = \pi / Asin(\frac{1}{2}.\ell/R)$

and if $\ell = R$:

$n = \pi / Asin(\frac{1}{2}) = \mathbf{6}$

Lorentz's Equation (magnetic force or field strength):

$F = q.v.B$

Which becomes:

$F = q.g.R.B$

for the laws of orbital motion

Where:

q is the total electrical charge = $q_1.q_2 / m_e.(q_1+q_2)$

v = relative velocity (electrical circuits)

g = gravitational attraction between m_1 & m_2

R = radial separation between m_1 & m_2

$B = \mu_o.I / 2.\pi.R$ kg/C {R = 2.R_n}

$I = e$

$B = \mu_o.e / 4.\pi.R_n = \cancel{4.\pi.R_n}.m_e/e^2 \cdot e / \cancel{4.\pi.R_n} = m_e/e = 1/RC$ kg/C

RC_e is the relative atomic charge of an electron {C/kg}

$B = 1/RC = 5.685634367312E\text{-}12$ kg/C

A6 The Heroes

The heroes of this story, to which I offer my gratitude, are listed below

It is not necessary to identify the invaluable contributions made by each of these contributors, they are all widely known and available in almost every scientific publication in circulation today.

Nicolaus Copernicus (Polish) 1473-1543
William Gilbert (English) 1544-1603
Tyco Brahe (Danish) 1546-1601
Galileo Galilei (Italian) 1564-1642
Johannes Kepler (German) 1571-1630
Christiaan Huygens (Dutch) 1629-1695
Isaac Newton (English) 1642-1727
Edmund Halley (English) 1656-1741
Charles-Augustin de Coulomb (French) 1736-1806
Hans Christian Ørsted (Danish) 1777-1851
Michael Faraday (English) 1791-1867
Josef Stefan (Austria) 1815-1863
James Clerk Maxwell (Scottish) 1831-1879
William Crookes (English) 1832-1919
Ludwig Boltzmann (Austria) 1844-1906
Hendrik Lorentz (Dutch) 1853-1928
Jules Henri Poincaré (French) 1854-1912
Johannes Robert Rydberg (Swedish) 1854-1919
Max Karl Ernst Ludwig Planck (German) 1858-1947

The others that were instrumental in the completion of this book are:

My long-suffering wife (Brigitte) sub-editor and critic

My daughter (Eléonore), who initiated this project

Kenneth Pickering friend & editor, who first suggested that I write it

My thanks go out to all the above each of whom have provided a valuable piece of the puzzle without which the final solution would not have been possible, along with my sincere apologies to anybody I have unintentionally omitted.